PASTE-UP

Production Techniques and New Applications

Rod van Uchelen

PRENTICE HALL PRESS
New York

First Prentice Hall Press Edition, 1986
Published by Simon & Schuster, Inc.
Gulf + Western Plaza
One Gulf + Western Plaza
New York, NY 10023

Originally published by Van Nostrand Reinhold Company, Inc.

Library of Congress Cataloging in Publication Data
Van Uchelen, Rod.
Paste-up: production techniques and new
applications.

Includes index.
1. Printing, Practical—Paste-up techniques.
I. Title.
Z246.V29 686.2′25 75-44584
ISBN 0-671-60830-4

Manufactured in the United States of America

10 9 8 7 6 5 4 3 2 1

Contents

Introduction

This book is designed to explain the functions and the mechanics of paste-up. It starts with its uses in basic office skills and continues with the more complex specialties, covering its role at three levels: office duplication, professional volume work, and art production. At each stage new applications are discussed and related graphic specialties surveyed.

New graphic techniques are constantly being developed as an outcome of new inventions in equipment and processes. These new techniques for graphic production, using the new equipment, may be contrasted with the formalized production procedures developed around traditional letterpress printing. Production is becoming increasingly involved with offset printing rather than letterpress reproduction. Related production techniques include computer-controlled photocomposition, strike-on typography, and transfer type as opposed to type set in metal. These processes are based on electronic and photographic means of image production, using the computer and the cathode-ray tube (CRT). The outcome of these technological breakthroughs is considerable change in production methods and work areas within the graphic industry. Changes in procedure blur the lines of demarcation that have traditionally existed, creating new areas of work and new job categories. Offset printing, silk-screen printing, photography, the CRT, the computer, and data-storage-and-retrieval systems such as computer on microfilm (COM) all provide new forms of graphic production.

The processes of most immediate importance to paste-up are computer cold type and offset printing. Paste-up is the easiest and least expensive way to prepare art for offset printing. Consequently, as the use of offset printing increases, there is a continually increasing need for artists who do paste-up.

Fed by the increasing need for information, there is a continued expansion in the use of printed material. Office graphics are becoming more frequent due to the computer. New processes and equipment make graphic production possible in an office environment. Some information must be preserved or circulated among a number of individuals, which requires duplication.

The graphic industry is assimilating the new equipment and changing its methods. Many technical and business publications use the new equipment because of its economic advantages. In fact, the lower cost of production makes it possible for some of these information journals to be published.

Many larger publishers also find it more convenient to do paste-up, or paste makeup, so that jobs once the domain of the graphic trades are now available to the paste-up artist. The volume of paste-up work is increasing as the printing industry switches to offset printing and automated techniques.

Advertisers are also changing to paste-up. Many find cold type suited to their needs now that the process is being refined and specialized.

New suppliers are constantly developing, and they are using computer systems. This creates yet more fields for paste-up. Typography and computer composition is one specialized area, and paste makeup for printers is another. Each requires special knowledge, but each also uses the basic paste-up technique.

All these developments make use of paste-up. Paste-up can vary from the very simple to the very complex. It depends on the requirements for the graphic design, the system of production, and the degree of specialization. Business-use graphics can be simple, but as involvement with graphic processes increases, the work can become more complex.

Paste-up is a general skill, and there are few commercial artists who don't make use of it to some degree. As an art skill it requires specialization, an appreciation of design, and some knowledge of technical and photographic processes. With the new technical developments its applications are still increasing. To make fuller use of these new techniques, all you need to know is how to prepare art for reproduction—that is, how to do paste-up. It's like learning to use the typewriter.

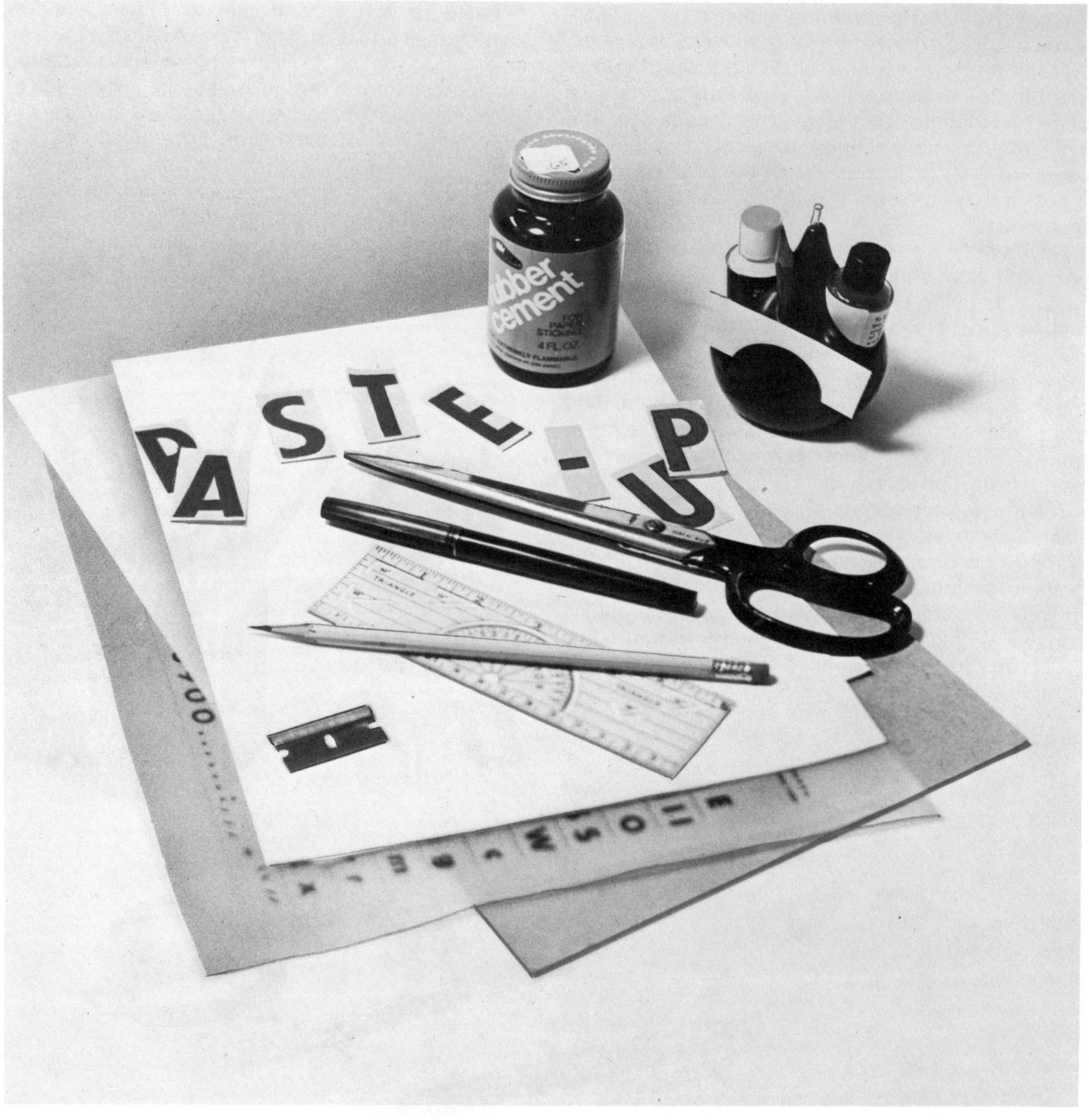

1-1. P-A-S-T-E—U-P spells out the job, and this is the simple equipment required for office use. You can probably find all of it in the average desk, except for the transfer type and the rubber cement.

1. General Office Uses

Preparing art for reproduction may seem far removed from office skills, but the first time you have a business letter that has to be sent to fifty people, the connection becomes obvious. How would you do this? Would you type fifty letters? Would you make a letter with fifty carbon copies? That would be a neat trick! No, you would use some duplication process.

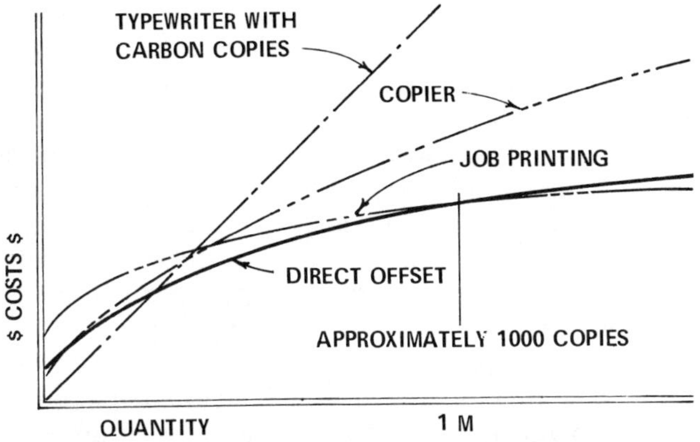

1-2. Job printing takes over from direct offset printing as the cheapest medium at about 1,000 copies. This figure can vary with the complexity of the job.

The economics are fairly simple. Labor is the biggest expense. Typing the letter ten times with five carbon copies would produce the fifty copies, but the fifth carbon wouldn't look too good, and don't make a mistake—that's five carbons. A copier such as a Xerox machine would cost less even if you prorated the cost of the machine and included it in the cost of the copies. With some reproducing processes you can even use office stationery (at extra cost), so the copies would look even better.

There are faster, less expensive processes than copy machines: the junior printing presses. The oldest is the fluid process, followed by the mimeograph, and now the offset duplicator. The offset duplicator is really a simplified printing press. These three processes all require a printing element. The fluid process requires a master; the mimeograph requires a stencil; and offset requires a plate. In each case the printing element can be prepared directly on the typewriter. It can be patched, corrected, and handled in a variety of ways. Its function is to *produce* the final copy, so it doesn't matter what it looks like as long as the result looks right. The cost for any of these processes, including preparation of the original on the typewriter, is low: the printing element costs pennies, and the time required to run off the copies is less than a fourth of the time needed for a copier.

Since business decisions are often based on cost considerations, the offset duplicator will undoubtedly gain in popularity. In addition to the cost factor, this process yields the cleanest copies.

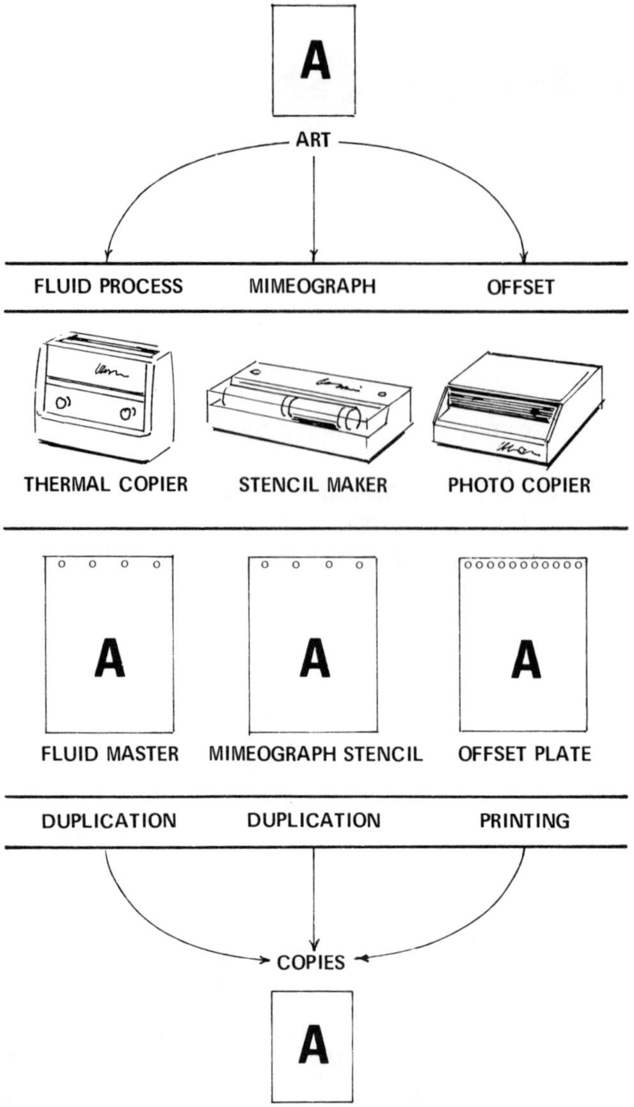

1-3. Each of the three printing processes can start with the same paste-up. A printing element is common to all, but the names and the quality of the reproduction change. Offset produces the highest quality, but the equipment is the most expensive.

There is another factor that has not yet been generally recognized: recently developed equipment makes it possible to produce the printing element for each of the three systems mechanically. The fluid-process master can be mechanically prepared on a thermal copy machine; the mimeograph stencil can be mechanically prepared on a thermal copier or by an electronic scanner; the offset plate can be prepared photographically.

Printing elements produced mechanically can reproduce anything that is black on white, even fine lines, coarsely screened photographs, and solid-black panels. This means that the processes are not limited to typed copy: the material can be handled as art and it can be done on paper in the format in which it will appear in the final result. Creating this format is called *paste-up*.

Office Duplicating

Doesn't the fact that you can assemble several pieces of material into a single letter suggest a number of different applications? You don't have to be letter-perfect in every detail because corrections can be patched in without showing in the final copy. You can also include a variety of elements such as charts or graphs.

All of these applications are possible with paste-up and a mechanical printing element. If your business letter has a logo or company heading—and what business letter doesn't—then your original has a total of three parts: the stationery heading, the body of the typewritten letter, and the signature. Each part can be positioned in just the right spot. While this can be done directly on the printing element, it is undoubtedly easier to work on a piece of paper and easier still to work with separate pieces of paper in the paste-up technique. You may wonder how you could reproduce the company heading directly on the printing element. The answer is that usually the printing would be on company stationery. By pasting up the elements and making the printing element mechanically, however, you could easily include the company logo. Copiers cannot utilize this kind of assembly because the edges of the separate elements cast shadows that show to some degree in the copies.

There are many other business uses for paste-up—forms that are distributed to numerous people or memos that always go to the same address and require the same heading, for instance. Printing copies of this repetitive information, with only the live matter to fill in, can cut the labor in half and reduce the costs too.

Office skills are largely involved with communicating and recording information graphically. Now, as more mechanical means of doing this are being developed, office skills might well start to include a knowledge of how to assemble the information for reproduction, that is, how to do paste-up for fluid-process, mimeograph, and offset duplication.

This simplicity of the paste-up process is what makes it so attractive. What you see is what you get. To do paste-up, you need the equipment to make the printing element, but the advantages of mechanical platemakers are easily seen in the speed and ease of operation and the evenness and accuracy of the resulting element. Mechanical platemakers vastly extend the range of what can be reproduced, and you can expect that if it is economical to have reproduction facilities in the office, it will also be economical to have the facilities for making the printing element. The equipment is not necessarily expensive in relation to other office expenditures.

Office Paste-up

The process of assembling elements to be reproduced on a piece of paper allows you to visualize what the finished result will look like. Depending on the fidelity of the reproduction system, the copy will look like the original. The main requirement for the systems of duplication under discussion is that the original be sharp, clean, and black.

The reproduction system responds to the black and to the light-reflective white surface. Any material that has this quality can be reproduced—a newspaper clipping, a drawing in black ink, or possibly a picture from a company brochure. This potential diversity may suggest a different approach to presenting information, a change from more conventional arrangements. Be careful, however, not to violate copyrights. If the material has been copyrighted, you may not reproduce more than a small portion of it, and you should credit the source even for that portion, unless you have permission from the original producer to use the material.

The material that is most often reproduced is typewritten copy. For best results keep the keys clean and, if at all possible, use a carbon ribbon—it is blacker and yields a sharper image than a fabric ribbon. Likewise, a pen will produce good copy more readily than a soft pencil, although a pencil is good for certain effects.

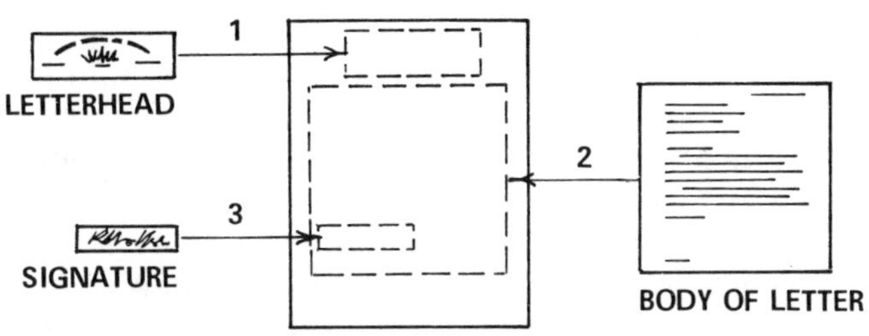

1-4. A business letter may suggest more complex situations for office paste-up. It is very convenient to be able to respace elements without having to retype the entire letter.

ACCEPTABLE	Typewriter (Nylon Ribbon)
GOOD	Typewriter (Carbon Ribbon)
BETTER	Proportional Typewriter (Carbon Ribbon)
BEST	STRIKE-ON COMPOSER (Carbon Ribbon)

1-5. Comparison makes the differences in quality apparent. If you must use a typewriter, the even striking pressure of an electric helps.

A pencil produces a line that is more gray than black, and, while the human eye can resolve this value of gray, the printing element cannot: the line must become black or white. That is, it will either print black or not register at all. The eye can respond pleasantly to gray effects, but the printing element can't. Reproduction will either become jarringly black or disappear completely. To get around these vagaries in reproduction, work with a definite black ink: what you will get is what you see. Stay away from grays and colors, which can shift out of the reproduction range.

It is sometimes convenient to use nonreproducing materials to "shift a mistake out of range." This can be done with white paint or by pasting a correction over the mistake. What would ordinarily be unsightly in an original the process can't "see," and the patch will not be reproduced.

The process of paste-up allows you to arrange the parts in the most pleasing position. You can include drawings sold for reproduction, transfer letters, rubdown letters, and other artist's aids found in art and stationery stores. As you select and arrange the material, you produce a design, hopefully more attractive and more effective for communication than typewritten copy alone.

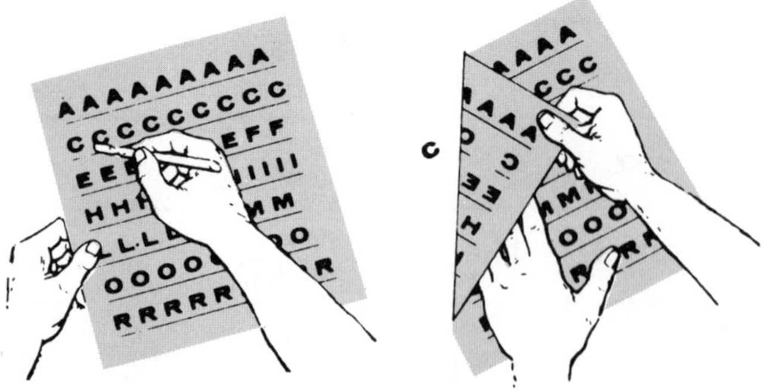

1-6. This drawing shows how transfer type works. The skill comes in the spacing and the design. (Illustration courtesy of Instantype.)

The next step is to adhere the material to the backing sheet. The easiest way to mount separate elements (without more specialized equipment) is with a small can of rubber cement with a brush applicator. This material will be familiar to anyone who has ever built models; otherwise you might think of it as a large bottle of nail polish—it smells about the same.

Spread the rubber cement on the back of the piece to be mounted, then drop this piece in the proper position. You can adjust the piece by sliding it around until the rubber cement dries. If any excess rubber cement shows around the edges, it must be removed, because it will collect dirt and the dirt will reproduce. The reason for using rubber cement is that the excess is so easy to remove. Just rub gently with your finger after it dries, and it will roll up and roll away. An even better method is to use a small glob of dried rubber cement from the edges of the can. Just rub the excess gently—it will adhere to the glob you are holding, leaving the paper sparkling clean for reproduction. Simple!

Cleanliness of work to be reproduced is important. Smudges, fingerprints, construction lines, or specks of dirt will reproduce if the tone is dark enough for the process to register. Dirt that collects at the edges of tape will reproduce, for example.

Paste and glue are more difficult to use than rubber cement, and generally the moisture will distort the piece of paper you are putting in place. Double-sided tape is better because it is hidden underneath the piece, but be careful of the edges, because they can be snagged up and damage the art.

What if you have to move something after it has been stuck down? It's not too hard to remove rubber cement—just use a few drops of rubber-cement thinner to loosen the bond, lift, reapply a fresh coat of rubber cement after the loosened piece has dried, and start over again. Something taped in place? That's a little more difficult—another good reason to use rubber cement.

Instant Printing

When the volume of office duplication is not sufficient to support the cost of the equipment, businesses often use an instant printer. The reception given to this recent office-services industry is attested to by its success and rapid growth. In most business sections several are available.

Instant printers use the same offset duplicator as is used in offices. It is a small offset press with few complexities and few adjustments. The major difference between this system, called *instant printing*, and offset printing performed by job-printing houses is that the instant printer produces the printing plate directly from the copy or art, while the job printer first makes film negatives of the art and then makes the offset plate from these negatives. Film negatives allow greater fidelity and quality of reproduction, and they can be manipulated in a variety of ways.

1-7.

1. Instant printing is available almost everywhere. PIP (Postal Instant Press) is the leading national chain of franchised dealers. This is the exterior of the Bloom franchise.

2. Inside the shop business hums: "You want how many copies by when?"

3. This is about a day's work ready for pickup at 3:45 P.M.

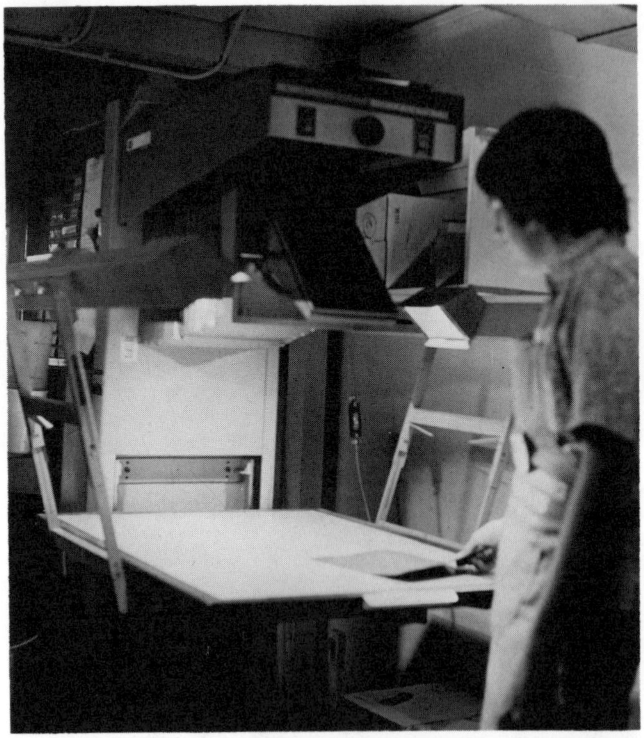

4. The copy camera used by the PIP chain can enlarge and reduce the original and is used to make paper plates.

5. The small offset presses run quickly: 500 copies in about 15 minutes.

6. The PIP chain uses this paper trimmer. It's small, but with 11 × 17 inches the maximum paper size it meets the requirements. For safety the black buttons at the extreme right and left must be held at the same time for the blade to operate.

7. The paper drill used by the PIP chain is similar in operation to a conventional drill press.

It seems natural to compare instant printing to job printing because they are both "printing," but direct-process offset should really be compared to fluid-process or mimeograph reproduction, because these are also direct processes. The quality of direct-process offset, or instant printing, has definitely improved, and this method produces an acceptable level of quality for most business printing, such as catalog sheets, small brochures, mailers, business communications, and office forms. Work that requires a higher level of quality such as color, slick sales folders, or large quantities is beyond the scope of the process, although it probably could and will be done.

Because instant printing is a direct process, the copy or art presented for reproduction must be complete. What you prepare is what the camera will see and what it will reproduce. Few adjustments are possible between the art and the reproduction.

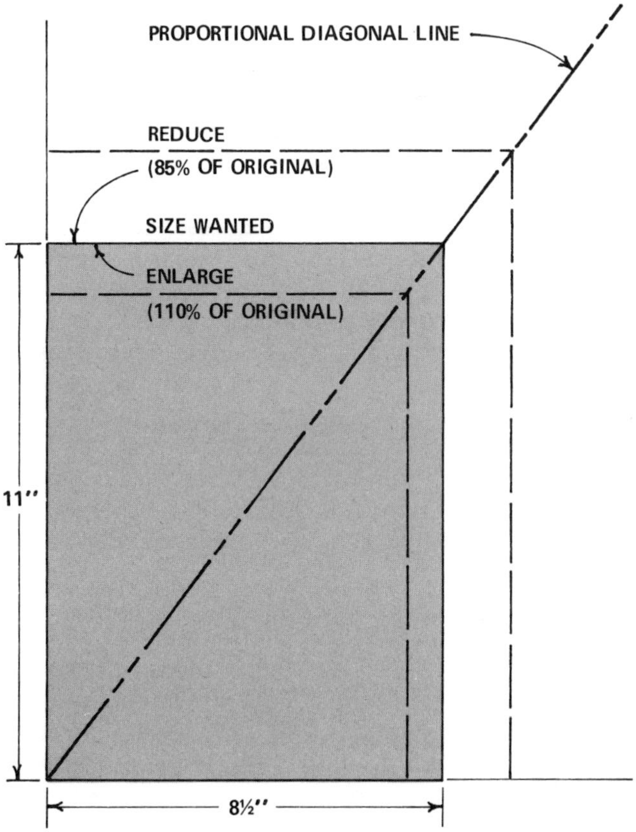

1-9. Many people who use instant printing don't realize that big problems can be solved with small size changes and that the quality can be enhanced too. Keeping the corner on the diagonal line maintains the proportions.

1-8. The negative used to make the plate is the major physical difference between instant printing and job printing, but superior materials and equipment also make a difference in quality.

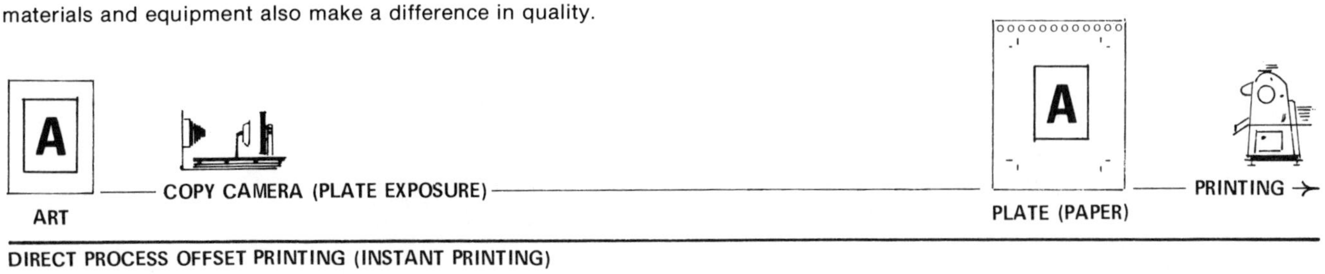

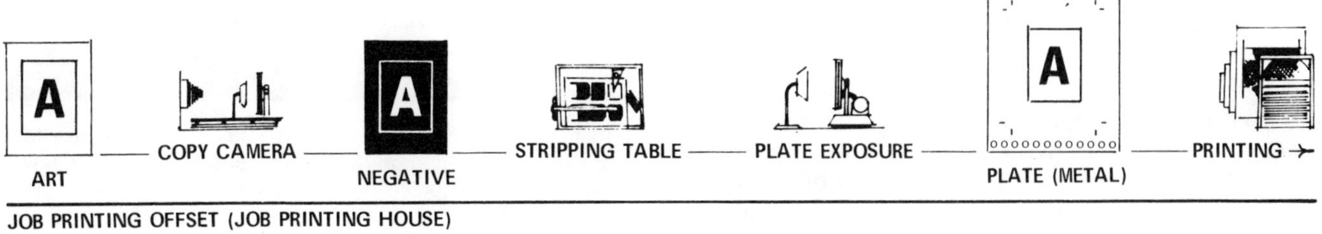

The process is systematized to work with standard sizes of papers and envelopes, just as with fluid process and mimeograph. The standard sizes of paper are stationery, 8½ × 11 inches; legal, 8½ × 14 inches; and double, 11 × 17 inches. The envelope used is the familiar #10 business size. Material is stocked in these sizes, and they are used for price lists. The 3-×-5-inch index-card size is often stocked, and the offset duplicating press is usually set up to feed only these sizes. This doesn't mean that you can't trim the finished piece to the size you need. Most instant printers are set up to do this. They can fold, gather, staple, punch, pad, or perform many other office services for a fee. As business users become familiar with these services and make use of them, quite sophisticated printed pieces will be produced with the instant-printing process.

Most instant printers stock more than two grades of white paper. They also stock colored papers, heavier paper, and cards. This will suggest many other uses of instant printing.

Because instant printing uses a camera to make the printing plate, most equipment allows size changes. The size of the reproduction can be increased to 110% of the original or reduced to as little as 50% of the original. This can be invaluable for fitting material into the standard-size format used for reproduction. It is generally best to reduce the size slightly, because this also reduces any irregularities and makes the reproduction appear sharper.

All the material produced by instant printing can be handled as paste-up. Many commercial artists use instant printers when the quantity and quality are suitable, and material produced this way is growing in acceptance every day. It *is* good enough.

The list of materials that can be produced with instant printing is almost endless. Many techniques used in commercial art are applicable to this process. It almost seems that strike-on cold type was developed just for instant printing—it wasn't. *Strike-on cold type* is produced on a machine that looks something like a typewriter, called a *composer.* The output is typesetting, and the type can be handled in the usual ways with justified margins and different typefaces. Transfer type also works well with instant printing, as do other artist's aids such as tone sheets. The results of instant printing can look very professional. Drawings and clippings can be included as illustrative material, but do be careful about copyrights. The instant printer who reproduces government documents or copyrighted material is subject to penalties. Not to labor the point, but a few copies are *not* okay. One copy of a government document is forgery.

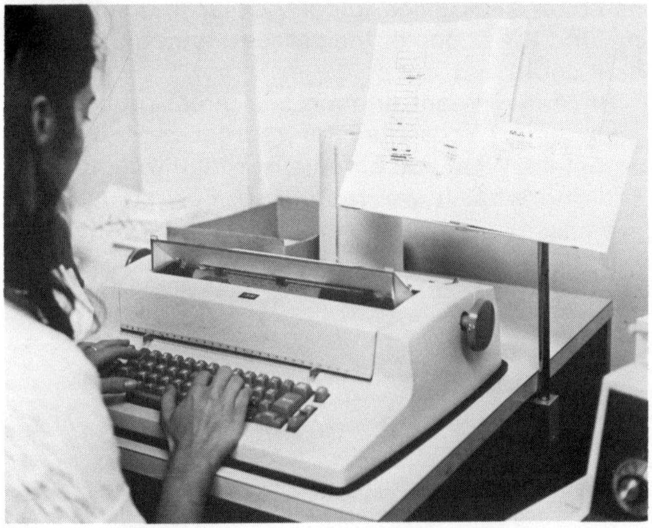

1-10 and 1-11. Located just above the Bloom PIP franchise, MJL II provides typesetting on IBM equipment. This efficient, low-cost ($5,000 to $15,000) system is just the thing for instant printing. MJL II also provides a mailing service.

Now You Do It!

The best way to learn anything is by doing it, and the best way to explain anything is by example. Pick a project for instant printing and run through this example. Then do your project in the way that you want, using the example for the information you need.

Many things can be done with instant printing, ranging from very simple to very complicated. This example lies somewhere in the middle—an announcement. It can include artwork, transfer type, cold type, and the basic paste-up technique. If you don't have your own materials, use the clip sheet supplied at the end of this chapter for your project.

Suppose that there is going to be a company picnic. The purpose of the announcement is to let all the company personnel know and at the same time to provide the necessary information about what, when, where, how, and how much. You have the information, and there are thirty locations where the announcement will be posted, plus fifteen to be mailed. To accommodate the mailing and the size of the bulletin boards, the announcement will be the standard 8½-×-11-inch size. The example will use white paper, but the color could be anything you like if it is available.

The first step is to plan the design. The easiest way to do this is to make a little drawing or diagram of what you want to do with pencil and paper. Plan how to handle the various parts of the message, what to emphasize, and how much room to allow. Think about how to prepare the elements—any little drawings, transfer type, portions to be typewritten—and how to arrange them. Normally the bulk of the information would be presented in typewritten form, but, in order to familiarize you with what a composer can do, this example will use cold type.

It doesn't matter if you change your original plan as you develop the paste-up elements for the design. It is better to stay flexible—to "wing it"—until you have everything in front of you. This is the moment of truth when the decisions will have to be made. Don't be surprised if the typewritten information takes up more space than you thought it would, for instance, or if the heading isn't as big as you thought it would be. Just adapt the paste-up as you go along.

For a small job it is best to use as much available material as you can. In this example "Manufacturing Company" was found in business literature. "You are invited to a" was produced with transfer type. "PICNIC" is made up of letters from a newspaper headline. The drawing is done with a felt-tip pen. The main body of the type is strike-on cold type, but it could just as easily be done on a typewriter.

Business printed matter is a good source for material, and there shouldn't be any copyright problem. The transfer type is available at art-supply or stationery stores. It comes in a variety of sizes and styles. If it is not available in a suitable size or style, it can be replaced with typewriter copy.

There shouldn't be any problem finding a newspaper headline, but it does take careful handling during the paste-up to keep the ink from smearing. Don't be tempted, while you're looking through the newspaper, to borrow a drawing from a food ad. It could easily create more problems than you realize. In many cases the art is from an art service. This art is copyrighted, and the service is paid for its use.

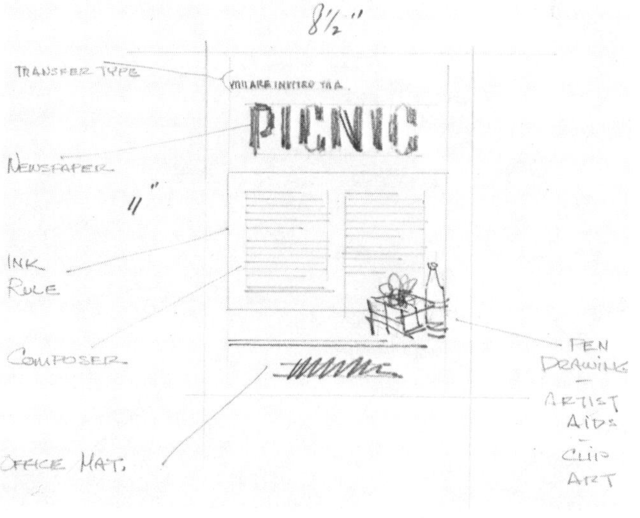

1-12. A small thumbnail drawing of the basic idea is a great help in planning the design. Good planning prevents problems from cropping up later.

They could charge you the entire fee for the service if you use their drawing, plus a fine. The same problem exists if a food-market chain produced the art, unless they give you permission to use it. You *can* make a tracing, however, and make your own drawing. How about splurging on a ready-made drawing, designed and sold for this purpose, at the art-supply store?

Type the main body of the message in the proper form and shape. In this example the copy was given to the cold-type operator for setting in the prescribed size and shape. Typesetting allows a larger volume of material to fit in a smaller space. You can draw the rule or line around the main portion of the message directly on the paste-up with ruler and pen. It is a good idea to put the letters for the word "PICNIC" on a separate piece of paper. If you want to change the position of the word in the final arrangement, you won't have to move each individual letter.

1-13. Notice how much less space typeset copy takes up compared to typewritten copy.

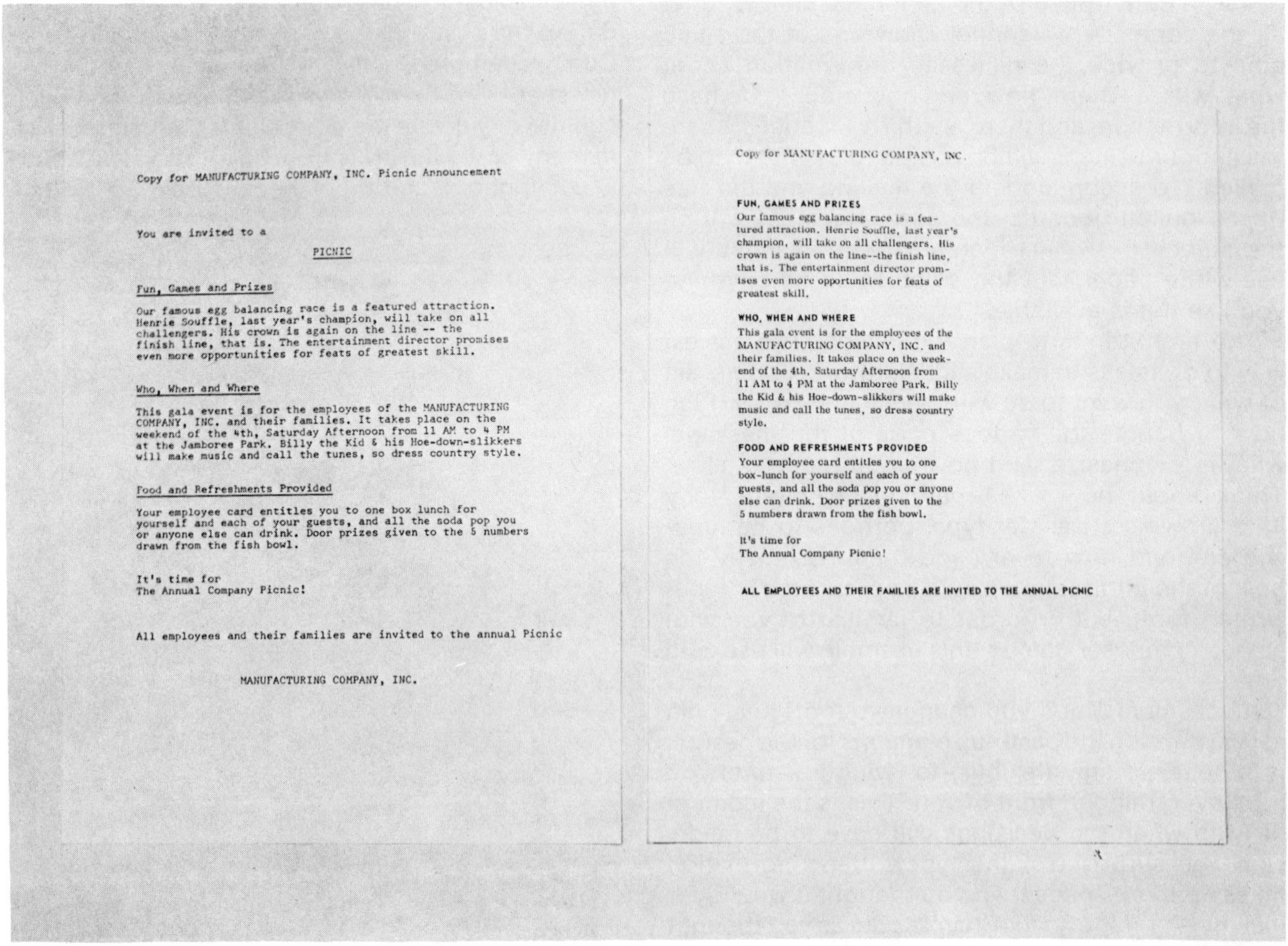

Copy for MANUFACTURING COMPANY, INC. Picnic Announcement

You are invited to a

PICNIC

Fun, Games and Prizes

Our famous egg balancing race is a featured attraction. Henrie Souffle, last year's champion, will take on all challengers. His crown is again on the line -- the finish line, that is. The entertainment director promises even more opportunities for feats of greatest skill.

Who, When and Where

This gala event is for the employees of the MANUFACTURING COMPANY, INC. and their families. It takes place on the weekend of the 4th, Saturday Afternoon from 11 AM to 4 PM at the Jamboree Park. Billy the Kid & his Hoe-down-slikkers will make music and call the tunes, so dress country style.

Food and Refreshments Provided

Your employee card entitles you to one box lunch for yourself and each of your guests, and all the soda pop you or anyone else can drink. Door prizes given to the 5 numbers drawn from the fish bowl.

It's time for
The Annual Company Picnic!

All employees and their families are invited to the annual Picnic

MANUFACTURING COMPANY, INC.

Copy for MANUFACTURING COMPANY, INC.

FUN, GAMES AND PRIZES
Our famous egg balancing race is a featured attraction. Henrie Souffle, last year's champion, will take on all challengers. His crown is again on the line--the finish line, that is. The entertainment director promises even more opportunities for feats of greatest skill.

WHO, WHEN AND WHERE
This gala event is for the employees of the MANUFACTURING COMPANY, INC. and their families. It takes place on the weekend of the 4th, Saturday Afternoon from 11 AM to 4 PM at the Jamboree Park. Billy the Kid & his Hoe-down-slikkers will make music and call the tunes, so dress country style.

FOOD AND REFRESHMENTS PROVIDED
Your employee card entitles you to one box-lunch for yourself and each of your guests, and all the soda pop you or anyone else can drink. Door prizes given to the 5 numbers drawn from the fish bowl.

It's time for
The Annual Company Picnic!

ALL EMPLOYEES AND THEIR FAMILIES ARE INVITED TO THE ANNUAL PICNIC

1-14.

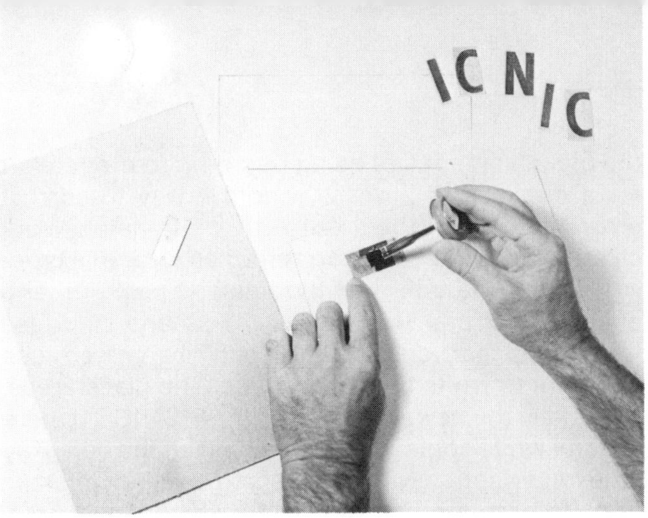

1. Cut out the letters, draw the line on the mounting flat, and apply the rubber cement, brushing past the edge onto the scrap paper underneath.

2. Repeat the process for each letter. The guideline helps you to align and space the letters.

3. With the word complete trim the mounting flat as with the individual letters, and this element is ready for assembly into the paste-up.

To do the paste-up, follow these steps along with the keyed illustrations.

1. Using scissors or a single-edged razor blade on a cutting board, cut out each letter. Draw a light line on the piece of paper you are going to mount the letters on to help you align them. Turn over the first letter, the "P," and apply a coating of rubber cement on the back. Use a piece of scrap paper under the letter to let you go past the edges and coat the entire back. Be careful not to get rubber cement on the face of the letter. Pick the letter up and drop it into position at the beginning of the line on the mounting paper.

2. As you position the rest of the letters, try to visualize the space for the entire word so that you can space the individual letters evenly. When the rubber cement is dry, remove the excess around each letter with a piece of dried rubber cement.

3. Trim the mounting, and the word is ready to be positioned in the paste-up.

4. To make the phrase "You are invited to a" with transfer type, draw a very light guideline on a separate piece of paper and rub each letter into position along the line. Trim the result, and this element is ready. Do the drawing in black ink on a piece of onionskin or thin bond paper that you can see through. It doesn't have to be great art—just something decorative. Cut the finished drawing out.

4. Use transfer type to make the lead-in. The process is similar to that for the word "PICNIC," except that the letters are smaller. Transfer type makes the letters easy to handle.

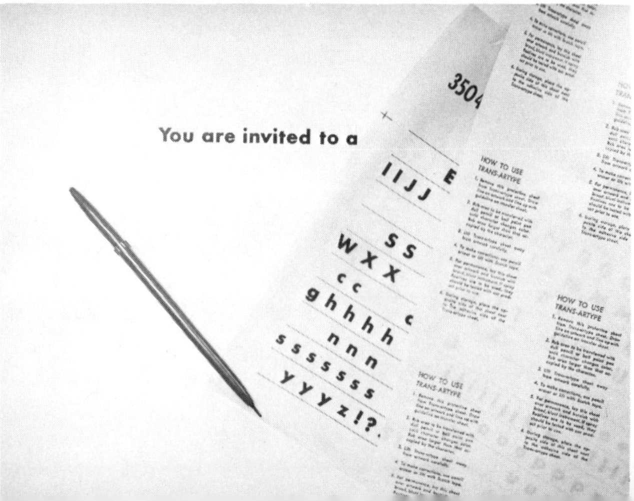

5. The basic principle is to cut each piece out separately for paste-up. When this is done, you are ready to arrange the pieces and to paste them into their proper positions. Lay all of the pieces on a piece of bond paper measuring 8½ × 11 inches and see how they fit. Change the design if necessary to achieve a pleasing appearance. With everything positioned to your liking, mark the arrangement and start pasting. Work the same way as with the letters for the word "PICNIC." Clean up as you go or, if you work cleanly, wait until you have everything in place. After all the pieces are pasted down, measure the points for the line around the body copy and draw it lightly in pencil. The measurements enable you to get the lines square to the edge of the sheet you are working on. The light pencil line shows you where to stop the ink line. If you are sure of the exact position, you can draw the line before you start the paste-up. Usually it is better to wait until you see where it belongs. Try not to make a mistake and bobble the ink line. But don't worry if you do make a mistake: you can always draw the ink line on another piece of paper and paste the new rule over the old one. This is one of the big advantages of using paste-up to prepare art for a duplicating process. Another way to correct errors is to paint them out with white paint. Work cleanly, as if you were correcting an error in a typewritten letter. Look over the finished paste-up and check for errors. Remove all marks and smudges that might reproduce. If they can't be erased, use the paint to white them out. Check the black areas, especially the newsprint letters in "PICNIC." If there are any white spots, black them in with the pen. Any little imperfection will reproduce. To make a mistake once is bad enough, but to duplicate it is worse. Take enough time with the original to make sure that it is ready for reproduction. Use the white paint and the black ink to make it a super job.

6. The paste-up should be complete and sparkling clean. To keep it that way, put it in a folder or put a protective sheet over it. It is surprising how many people don't understand that a paste-up has to be kept clean. If someone has to okay it, have them initial the back of the paste-up or the protective sheet. A paste-up takes time to produce, and in business time is money.

5. With all the elements completed the design problem is now solved. Follow the original thumbnail drawing, making adjustments if necessary. Attach each element as with the individual letters, and the paste-up is finished.

6. The paste-up with its protective folder. Notice that the elements are aligned exactly during the final mounting. The rule is drawn, the art retouched, and the letters blackened, and the paste-up is ready for reproduction.

There are still a few things to watch for. The art is photographed under a strong light to make the plate for instant printing, and it can cast a shadow at the edges of the paper. This shadow will reproduce as if it were a line. The instant printers watch out for this and to a certain extent they can correct the printing plate, but you should work with the printer. If two pieces of paper nearly butt together so that a shadow is cast between them, fill the gap with white paint or overlap the edges.

Make sure that the plate is properly exposed and that the prints are properly inked. Even with perfect copy there is margin for error. To be fair, however, you cannot expect the same results from instant printing as from offset printing with a metal plate.

1-16. Printing is not as automatic as it seems: considerable skill is involved in fine printing. In this example compare the thickness of the lines, particularly in the face, hands, and trumpet.

1-15. A summary of mistakes that can spoil the printed effect. The printer cannot deal with these since the paper plate is made directly from the art.

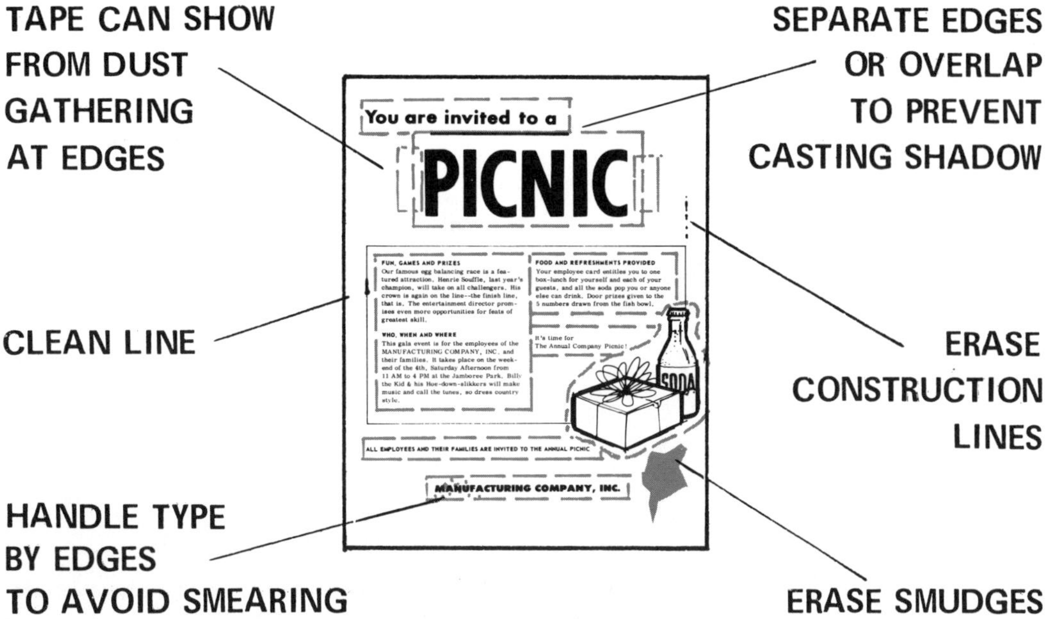

The design of this announcement could be modified in a number of ways. It could be done as a typewritten sheet, for instance, with the word "PICNIC" very large so that the notice would be seen on the bulletin board. Any of the elements could be left out or substituted for. The announcement could be made into a handlettered poster. The design depends on what materials are available and on what you can do yourself. This kind of project usually has both explanatory copy and decorative elements to add interest. Simpler is generally better. You might find it easier to simply type the message on a piece of paper in a suitable position and add a display heading. This would simplify the paste-up work to just the heading.

If you feel slightly clumsy and confused, the techniques are clarified in succeeding chapters. You don't have to be an artist to do paste-up, but the same principles used in commercial art apply to work prepared for instant printing and office duplication. How skilled you become depends on your aptitude and your desire to develop your abilities. It's easy to learn *how* to do paste-up, but to be good at it you have to *do* it, to practice.

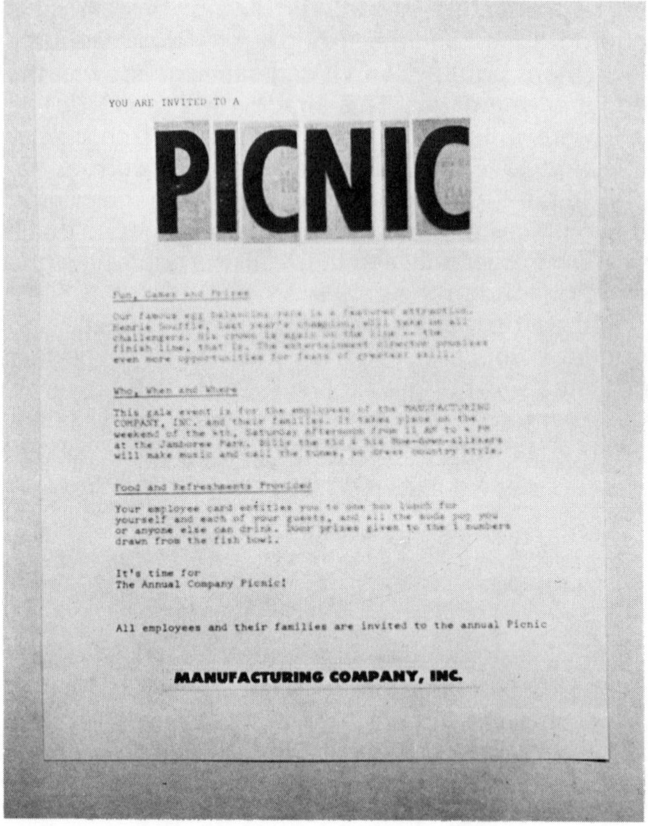

1-17. A design variation that would be easier to do and yet would fulfill the purpose—it's a matter of judgment.

1-18. This clip sheet is duplicated at the end of the book in case you make a mistake or for additional practice.

CLIP AND USE THIS PAGE FOR MATERIAL TO DO PASTE-UP

You are invited to a

CLIP IT CLEAN!

FUN, GAMES AND PRIZES

Our famous egg balancing race is a featured attraction. Henrie Souffle, last year's champion, will take on all challengers. His crown is again on the line--the finish line, that is. The entertainment director promises even more opportunities for feats of greatest skill.

WHO, WHEN AND WHERE

This gala event is for the employees of the MANUFACTURING COMPANY, INC. and their families. It takes place on the weekend of the 4th, Saturday Afternoon from 11 AM to 4 PM at the Jamboree Park. Billy the Kid & his Hoe-down-slikkers will make music and call the tunes, so dress country style.

FOOD AND REFRESHMENTS PROVIDED

Your employee card entitles you to one box-lunch for yourself and each of your guests, and all the soda pop you or anyone else can drink. Door prizes given to the 5 numbers drawn from the fish bowl.

It's time for
The Annual Company Picnic!

ALL EMPLOYEES AND THEIR FAMILIES ARE INVITED TO THE ANNUAL PICNIC

MANUFACTURING COMPANY, INC.

PASTE-UP

2-1. Though this concept is similar to the one in chapter 1, it presents a more complex problem and requires discrimination in handling. There are many ways in which it could be done.

2. The Graphic Industry

Paste-up has traditionally been a job for the artist. There are many situations that require considerable skill, and the greater the skill, the better the job. While some work only requires following a simple format, positions in which considerable skill and artistic judgment are needed are still jobs for the artist.

The degree of difficulty that a paste-up can present varies widely. In any given situation there are numerous ways in which a paste-up can be done. To use the announcement from chapter 1 as an example, what if the invitation to a picnic were an invitation to see a movie entitled *PICNIC?* Even if the same design elements were included, imagine how differently they would have been developed! How much more complex the job could be!

The word "PICNIC" might be part of the illustration. Say the lettering was done by one artist and the illustration by another, both were different sizes, and neither was the right size for the paste-up. Suppose that still more information had to be included in the body type, that there were two kinds of type in different sizes, that the rule was ornamental, and that the line at the bottom was a complicated producer's logo—something like Bright Star Libido Productions! For the paste-up artist this job would be a real picnic!

For years paste-up has also been a position for the schooled apprentice. Under the guidance of more experienced professionals he or she put together designs and learned the mechanics of the profession. He brought to the job the training and ability developed in art school, but he still had to learn a great deal about the business of commercial art, starting with the seemingly low levels of errands, mounting, and matting art and eventually including dealing with suppliers, handling art, and producing complex designs.

The paste-up artist did not start at the bottom: he *was* a trained artist, but because he needed to develop commercial judgment, he apprenticed to develop this skill. Art schools tend to leave these practical aspects of training to on-the-job experience because they have to give the student a basis in design. Some art schools also require academic prerequisites because there is scarcely enough time to teach the fundamentals of such a vast and complex field.

Paste-up became the professional training ground, perhaps because the work could be supervised. It was also the area to apprentice talented newcomers with no training. This was how things were done, but not any more. Paste-up has changed and is changing still. It is a bigger and still growing field within the graphic industry. There are more and more calls for paste-up work, and the work has a new and broader meaning. There are new applications for which no training is yet available, and still more are developing. Art schools are paying more attention to teaching paste-up.

These new areas are a result of changing techniques in business, publishing, printing, typesetting, and the more traditionally oriented field of advertising. Paste-up is being used with data filing and manipulation of computer and film (COM) in industry for catalogs and publications.

These are all jobs for the artist, and that is how they are advertised—"Paste-up artist wanted." But paste-up has a new and different meaning—art technique and design ability are needed in relation to other technical skills. Often there is no such marriage of skills in what were formerly different fields of expertise. The jobs go unfilled, but talented people pick up on these new combinations of fields. The basic skill in these areas is paste-up.

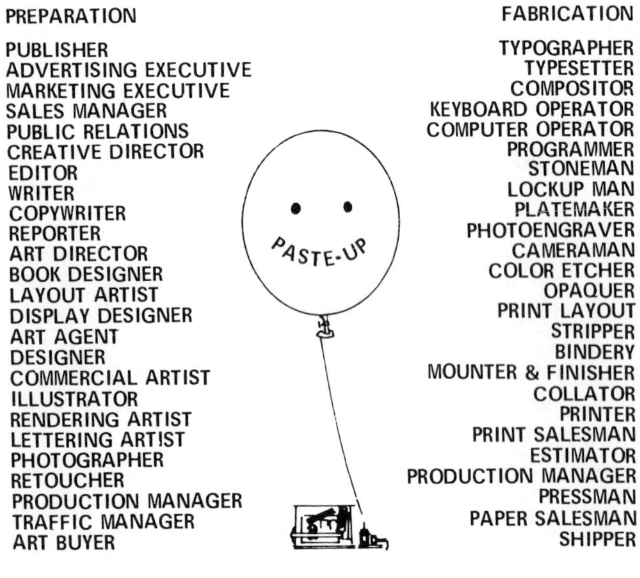

2-2. A happy face is just the thing for dealing with so many people. This is a partial list of the personnel a paste-up artist may work with.

Paste-up Specialties

The position of the paste-up artist in the graphic industry is that of a producer: he or she actually *produces* the finished product. Paste-up is the one essential that can't be skipped, and sometimes it is the only graphic work produced between concept and printing the finished product. Again, paste-up is often the last step in a long chain of command. There might be business managers, creative directors, art directors, editors, production managers, artists, photographers, designers, and more—the paste-up artist is handling their work. At the other end there are printers, typesetters, and a host of other suppliers—the paste-up artist is preparing their work.

As you have seen from the previous chapter, the paste-up artist may be the *only* artist *and* the person who deals with the printer. More often, however, he deals with a great number of people within the graphic industry.

One paste-up specialty that is rapidly developing is newspaper paste makeup. Cold type and offset printing are now used to produce newspapers; they are no longer composed as metal and locked into page forms. The pages are put together with paste-up, and the process is termed *paste makeup*. With this changeover newspapers are dispensing with a portion of their staff and hiring additional new people who can do paste makeup. Personnel are hard to find, and the papers usually have to hire talented beginners and train them to work with their typesetting equipment. This process is relatively new, and on an industry-wide basis occupational categories have not been formalized. Paste makeup requires skill in handling typographic material and in doing paste-up rapidly.

The Range of Paste-up

Between traditional paste-up and the new applications that are developing along with new technology are many changes in the range of paste-up both *within* the graphic industry and *by* the industry to develop talent. There are varying degrees of difficulty in the work, need for design ability, and need for technical expertise.

There is no one correct way to do paste-up: it is subject to interpretation, and there are many different opinions on the best way to handle art. It becomes even more complicated if the requirements for quality are coupled with or replaced by requirements for speed. If you add still another factor, several different production systems, you will realize that paste-up is anything but cut and paste.

Decisions on the requirement of paste-up for high quality, for example, are based on availability of equipment, abilities of personnel, mechanical capacity of the printing press, skill of the printer, and other factors that vary in each instance. All of these factors must be balanced and a decision made. This would never be left to a beginner; someone with more experience makes these decisions. But the beginning paste-up artist may still have to select the technique best suited to the requirements of the project.

2-3.

1. One of the jobs for the paste-up artist is correcting typographical errors. This mistake in galleys is corrected by placing the new type over the old and cutting through both galleys simultaneously to make a mortise, or opening, for the correction.

2. With the mortise made and the galley backed with tape, the correct type is dropped into the exact position.

Paste-up is as difficult as the application. Consider a newspaper page, for example: it basically requires only handling columns of type to execute. Provided you have mastered the skills, it shouldn't be too difficult. But the front page, with its effect on sales, merits the attention of the city desk and an editorial meeting. The arrangement is important—the right stories have to be in the right place with the right display.

If there is more to pasting up the front page than just paste-up, maybe the inside pages would be easier. This is true until a late ad comes in, which may require tearing up a completed page, reworking it, killing part of a story, and fitting in the ad—best of all, you have 15 minutes to keep the press schedule. Assuming that the editor is angry, the ad manager is insistent, the foreman is upset, and the lead pressman is shouting for more plates—still easy?

This example of a supposedly simple job raises the question of what artwork is. Matisse provides an example. In his advanced years he cut colored paper into forms and designs and pasted them into delightful compositions. Anyone moved by art would think that, whatever the means, Matisse's understanding and art were expressed to the fullest.

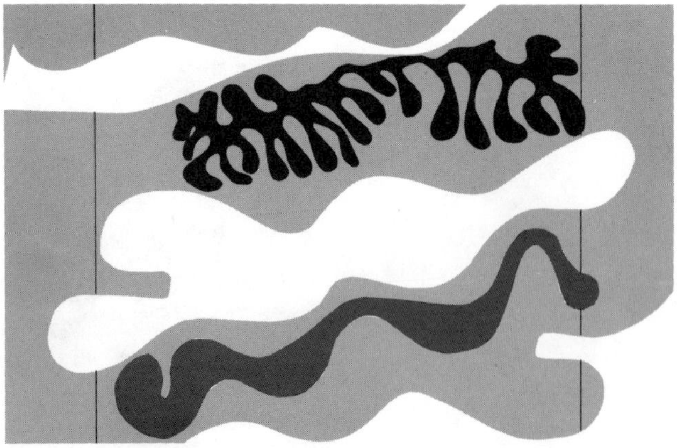

2-4. Matisse's paper cutouts have a precedent in Teriade's *Jazz* series (1944–47), as shown in this copy of one of the illustrations. Bright colors enhance the visual movement.

This example makes the point that good design and sensitivity to composition are important to paste-up, whatever the project. You must always consider the subjective purpose that the work performs.

Matisse used paper cutouts for artwork. Renoir, temporarily lacking the use of his arms, painted with his feet. Paste-up can be very mechanical. Is it art? It can be, depending on how well it is done. What makes craft into art is excellence. Paste-up certainly can be artwork, and it should be handled as such by the artist.

There is little question that skillful paste-up of complex designs is an art. It is the more mundane, mechanical designs that cause issue. Artists realize, however, that almost any graphic work can be handled as art, which may be why they almost invariably look first, read last—nature's way of protecting the artist from some of the content of paste-up work. Never mind if it makes sense—how does it look? What it says is the copy department's problem. This departmentalization is unfortunate, for this is how the corporate committee designs the horse that becomes a camel.

Ideally, the paste-up artist could catch the misspelling in the heading that is too obvious for the experts to see. He or she should understand the designer's idea and pull all the elements together in just the right way. This kind of paste-up soon becomes design as well. Art is assimilated as you go along. It can take a little time, since you must develop an understanding of the subject that is involved in a commercial job as well as design ability.

Paste-up is used by more businesses than ever before. Newspapers are a new field, as are typographers. Publishers of books and magazines use paste-up to produce their work, and there are more and more publishers. Advertising agencies have always used some paste-up to produce collateral material for clients, but this material is often handled by the advertiser himself nowadays. Industries that need to produce parts catalogs and industrial-instruction brochures, such as aircraft companies, generate a considerable amount of material. Some of it is handled by large art departments and beautifully executed. Other areas for paste-up are internal business management, such as sales material or house publications, and mail-order, a rapidly growing business.

The requirements for work in an advertising agency will probably be different than the requirements for work in a publishing house, for example. The nature of the work in any job depends on the material that is being worked with. It is hard to think of a business that doesn't use some printed material. Usually, however, larger organizations have more work for a paste-up person who needs experience than smaller ones. These larger organizations tend to step the work more than smaller art departments, in which each artist is expected to do several jobs.

Preparation for Paste-up

Full-time paste-up, in addition to artistic talent, requires the same abilities as any other office work. It relates to the work of many other people, to a volume of detail, and to an ability to organize yourself and your work. Although there are environments in which paste-up is used, some of the qualities that are needed for this type of work might be generalized.

People are important. Paste-up brings you into contact with all kinds of people with varying dispositions and under different amounts of pressure. Your working environment can be anything from the printing plant of a small-town newspaper to a plush advertising office. The attitudes of fellow workers can be anything from radical press to blue-serge bank.

Wherever you work, an ability to organize the work flow is useful if not essential. Your art ability should distinguish the work that *you* produce. You will probably have to deal with other artists and suppliers.

If you work in a department-store advertising department, for instance, you will be involved with layout artists, whose work you will handle, and with the art director. You will also come in contact with copywriters and buyers, who are interested in how their merchandise is displayed. At later stages there are newspaper reps, the platemaker, and others who produce the work. With a number of ads scheduled for the day and many pieces shown in each, you will be faced with a lot of questions—"Where is the. . . . ?", "Did you get the art yet?"—as artists and copywriters track down the progress of a particular job. The newspaper will want to know about any changes in art; the platemaker may want clearer instructions on what to do with a particular piece of art. An organized system of work flow helps prevent midnight phone calls, missing artwork, or last-minute redos for an imminent deadline.

At the other end of the spectrum is the art studio, probably too demanding a spot for a beginner. It requires the broadest range of paste-up skills because of the variety of work that is encountered and the demand for design skill. Depending on its accounts, you could be called on to do almost any kind of paste-up, including brochures and catalogs, which have many more pieces than even a complex ad. A paste-up artist in this position has to keep track of the art as it comes in from an individual artist and as it goes out for processing. He shouldn't have to track back to the artist if missing art is sitting at the stat house.

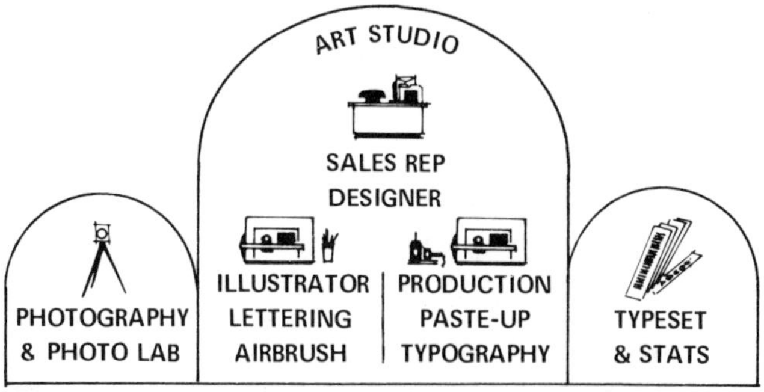

2-5. An art studio is an excellent place for the beginning paste-up artist because of the variety of work, but unless the studio is very large, there might be too much variety for the beginner to handle.

Some people have the knack of keeping an enormous amount of detail in their heads. Others devise checkoff charts for work flow. There are a number of methods. Organization doesn't really have anything to do with paste-up per se, but it is essential for doing the work in a business situation. Organizational abilities are often just as important as the actual paste-up work. They involve keeping track of the job you are handling, office procedure where you are working, and your background knowledge of business practices in general.

For best results with the actual paste-up work, the professional should approach each job as a fresh assignment. There are many areas for improvement, such as your speed, cleanliness, accuracy, problem-solving ability, and design skills. Every job presents a fresh set of requirements, and the professional can take a fresh approach each time.

In big shops there is usually more specialization. It increases efficiency with a large volume of work, especially if the nature of the work is somewhat similar. If the jobs are very diverse, the work could not be systematized, but systematization does allow different individuals to do what they do best.

In order to relate your own work to the work of others, it is very often helpful to take a guided tour to see what happens to what you do. For example, after you have prepared a paste-up, follow the work to the printing plant to see how the cameraman and the photo stripper handle it. It's fun to see the printing presses in operation—you're involved.

Even with all these professionals working on your design, it can't be any better than it was on the paste-up. This is what the cameraman photographs to make the negatives and what the photo stripper mounts to make the plates and what is duplicated on the printing presses. Your contribution is the aesthetics that are incorporated in the paste-up.

If you have the aptitude, you can learn paste-up on the job. But it is easy to see that it would be easier and that you would progress faster if you had art-school training in design and drawing. Almost anyone can be taught to draw. Art schools develop the aesthetic sense used in design and teach the mechanics of art.

You can prepare for work in the graphic industry while on the job by developing your abilities. Remember too that art is always done in some particular context. General knowledge will give the flexibility to work with whatever material comes along. A given job often requires expertise in a particular subject—think of an editor, a journalism major who works on a business publication. After years of dealing with business subjects, the editor will surely develop a keen sense of them. Starting out as an expert in journalism, he or she will almost certainly become an expert in business as well.

With the new generalized application of paste-up in business via computer typesetting and offset printing, paste-up techniques can be compared to the more usual office skills. Instead of thinking of paste-up strictly as an art skill developed by an artist, the technique is so useful for basic office use that a lesser level of skill is almost essential for office duplication. The difference between the two is determined by the application and the amount of specialized knowledge required in addition to the paste-up.

Paste-up can be office work. It can also be very delicate artwork. To specialize in paste-up—to make it an art job—you should evaluate your aptitude and your desire to go on to an appropriate career in graphic arts.

Paste-up and Production

The professional, commercial paste-up artist is engaged in what is called *production*. While paste-up can be used as a design technique, the more usual use is for production, which is generally specialized in medium-to-large-size art departments.

Artists who specialize in production handle paste-up and related functions involved with developing an art director's or a layout artist's design. In some instances they may even select illustrators, photographers, and artwork. They usually handle type specifications and contact with suppliers. In short, a professional production artist might handle everything connected with producing a given design. In a larger art department the specialization can be very fine: some production artists handle nothing but type, for instance.

It is not hard to see why an artist needs to know about production. If you are an illustrator, for example, it would be hard to make intelligent decisions about tone or color if you don't know how an illustration is produced. The fact that you do illustrations doesn't change your need to know. Large art departments create this kind of specialization; an art department where you can pick up general information is a great place for a beginner.

Some artists, after gaining their initial experience, move to a better situation in a smaller art department. Here they will have a broader range of assignments and less specialization. There are even one-man art departments where the artist wears all the specialist hats. This is a job for an advanced professional. This artist may plan an advertisement together with the executive staff, design it, handle changes and approvals, buy the art or do it himself—changing to the illustrator's hat—and then—changing hats again—do all the production up to delivering the finished paste-up to the printer. Such a multiplicity of talents usually requires many years of experience, for it involves client contact, pricing, buying, bidding, and the ability to design, render, illustrate, plan, and handle all the details of production as well as doing the paste-up.

Someone in such a position may well skip certain steps in the development of a particular job. He can afford to because he has no coordination problem and can visualize to his own satisfaction the steps that he is leaving out.

This may be deceptive to you if you are a novice in graphics, because you are apt to think that the omitted steps aren't necessary without realizing that they have in fact taken place mentally. It can lead to doing design as paste-up—combining the design function with the paste-up—which is fine if you know what you are doing and can imagine variations in concept and design, but if not, you will be amazed at what you're missing.

One innovation that is being seen more frequently—along with the new techniques of cold type—is the use of a design format for production, either prestyled or the result of trial and error.

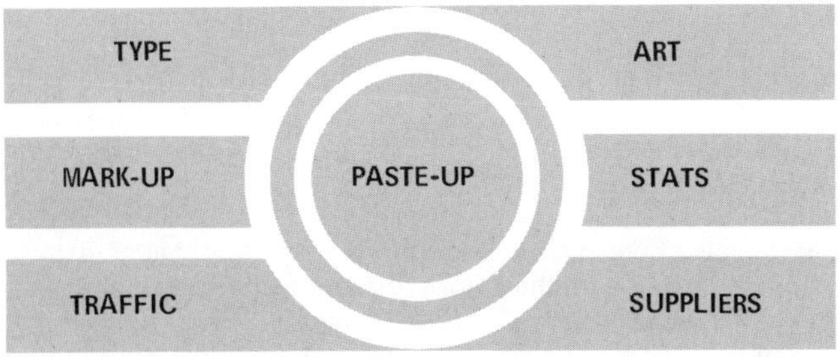

2-6. A generic view of the usual paste-up duties.

2-7. The top drawing illustrates the traditional development of graphic material, but the other two systems show new trends. They are important for the paste-up artist, because they create more responsibility. The center system works best with experienced, cooperative people and may account in part for some of the graphically simple designs on the market and for the switch some publications have made to a tabloid format. The bottom system attempts to eliminate the design function entirely.

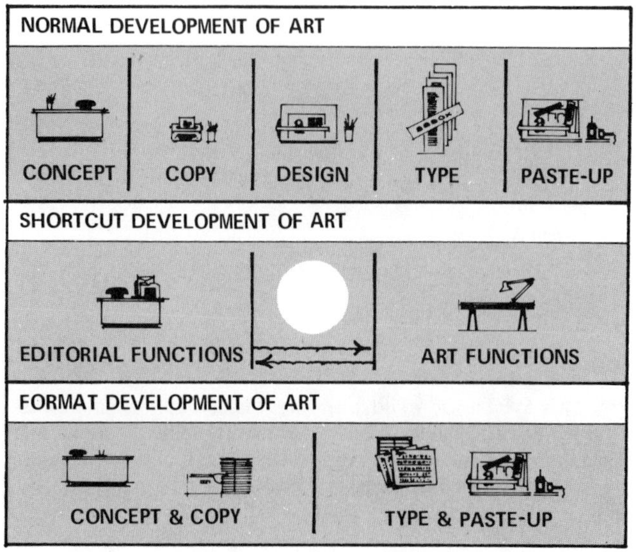

The production work is done to certain specifications or within prescribed limitations, and there is no actual layout or blueprint to work by. The sizes and forms for the design are preset, and all that is required for paste-up is to arrange these elements: there is no actual design.

If the format is good, this can be a successful approach, and it certainly cuts down on the amount of labor involved. It can become a serious limitation, however, if it is continued for too long without a change.

To illustrate this, think of a newspaper in which repeated use of certain typefaces creates an identity. The format for presenting the information is also standardized, as is the typical use of pictures. Because of the great need for speed in producing a newspaper, there is a functional reason for this format, but if similar pictures were always in the same places with the same type, you would think you were reading yesterday's newspaper. Even a newspaper needs makeup within the format for interest and to convey the news more effectively.

How much more, then, do publications that deal with abstractions and a variety of ideas need a treatment that expresses the thought that is being communicated?

The necessity of creating variety within a format gives more responsibility to the paste-up artist and requires more design ability from him. Variety can be achieved, and there will be an increasing need for paste-up designers and artists capable of more than tasteful arrangement of the elements in a paste-up.

Any professional paste-up job involves some particular production system. Whether the paste-up is for publication, catalog, or advertising work makes quite a difference in the kind of production that is used.

The methods used for art production—and the aesthetic values considered important—vary with each project. In fact, different companies have different policies on handling art, which reflect management value judgments and decisions about the relative importance of quality, cost, and speed.

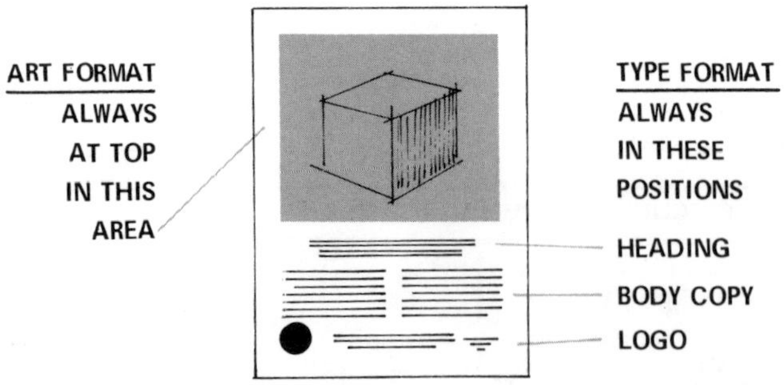

2-8. This format for an advertisement is sometimes called the Ayer format after N. W. Ayer of Ayer Advertising Agency fame. It is so basic and workable that it is still often used, and it shows what a format is. Formats can be defined in many different ways.

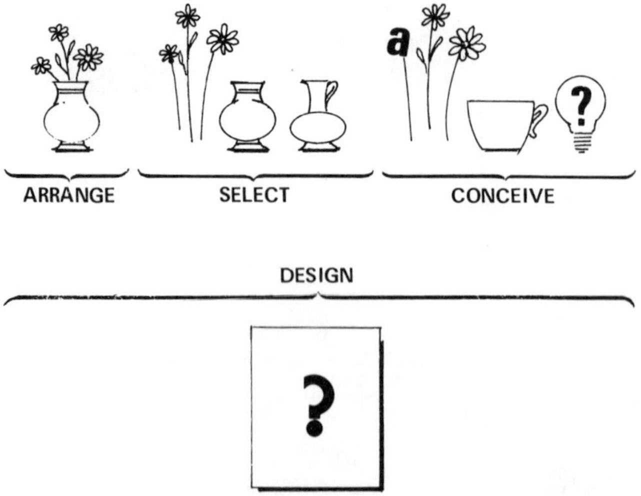

2-9. Levels of design approach are indicated by this concept diagram. Left to right they move in ascending order of difficulty, as should be easy to infer, because merely arranging elements is simpler than also deciding what they should be. These distinctions in design function are important because they limit or extend the boundaries for the paste-up artist.

It is impossible to generalize such policies, but you can expect advertising work to be more concerned with quality than catalog work, for example. This is so because advertising deals with public presentation, whereas catalogs may be strictly for inplant circulation. The requirements of a given piece of work dictate the production approach and also reflect the budget available for the job.

There are many ways in which a design can be done and the finished art and printing handled. For example, line art, as used in the example given in chapter 1, is less expensive than the continuous tone of a photograph; or, if photographs are processed within a company to reduce the unit cost, it may be a matter of policy to leave them unmounted. In this case it would be considered preferable to take the risk of damaging the art than to spend the time and money to protect each photograph with a mount and flap. On the other hand, if the job was an advertisement and the photograph represented a cumulative investment of about $500, the photo would most certainly be mounted, flapped, and protected!

Many production systems are developed by trial and error, and the policies arrived at are the result of experience. There are many ways to produce the artwork and the printing. The broader your knowledge and experience, the more choices you have. An experienced production manager knows where to get the best work, how to cut corners and gain speed, and which processes are the most and the least expensive. Production decisions balance the considerations of quality, cost, and speed, and the outcome is a production system.

You can see what makes the production expert an expert—but what if you had to prepare art for printing without a production system? The commonsense thing to do is to ask the printer or platemaker who is doing the job exactly how he needs the art. He or she is an expert in the area, and you will most likely have to talk to him about printing prices anyway. With these basics you will surely be able to get the job done.

If the job is supposed to be inexpensive, stick to black and white. As you move from line art to simple halftones to more complex treatments, you are also moving up in cost. Try to present a complete pasteup—the more work the platemaker has to do, the higher the cost will be. If the quality is suitable for the job, use the paper-plate and offset-duplicating system, but since the paper plate does not permit manipulation, the art must be complete and ready for shooting as line copy. Figure the costs, the quality, and the time required—you'll be well on your way to a production system!

New film and reproduction techniques are increasing production possibilities every day, and production systems are ripe for innovation. Production is becoming as creative as design, and since the two fields work together to achieve the desired results, you should have some understanding of and experience with both.

Once you understand the job you are doing, it is useful to keep track of the time spent in doing the various stages. Time is secondary compared to quality, but for business purposes it is often the basis of payment, especially for freelance work. If you don't know how fast a professional would do the work, measure yourself against your own best time, which should be an encouragement as your skills increase.

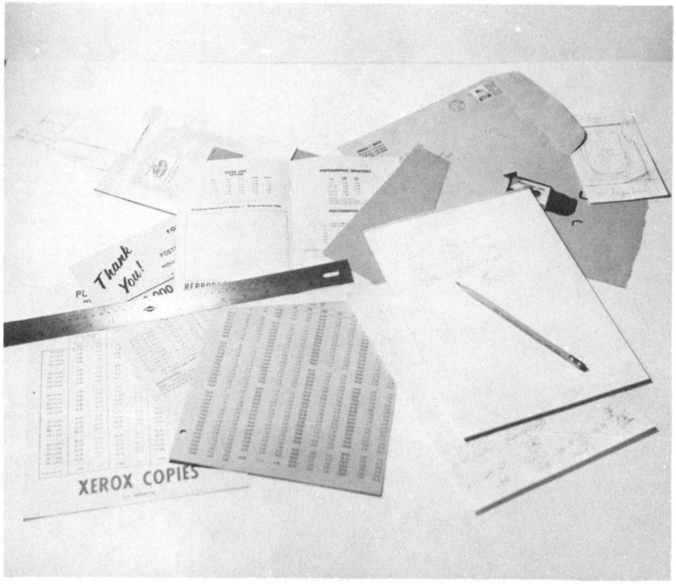

2-10. A production system is a balance of costs, methods, and expectations.

3. Materials, Tools, and Skills

Many skills are needed for paste-up, and the more you have, the better. These skills involve the use of various materials and tools. There is usually more than one way to do a specific thing, and how you choose to do it depends on which tools and materials you can use. The ability to use drafting instruments like the T-square and the triangle is necessary in order to make ink rules or to cut and mount type and art. More important than any of these skills is to understand why they are used, what preparation is required for art reproduction, and which steps to take to arrive at this result.

The most basic skill, then, is an understanding of how a given piece of art can be reproduced—what the choices are and how these methods of handling art will affect the reproduction. In other words, you should understand what steps to take, what tools are needed, and how to use them.

3-1. Compared with the array of tools for office paste-up, the accumulation of pens, pencils, and brushes on this taboret may seem unreal, but each has a special purpose in accomplishing a particular task in the shortest amount of time. You don't need all of this equipment at the beginning—it is acquired over a period of time.

It may take some time to develop judgment about the subtleties and tastes of the people you are working with and the techniques most often used in your office. Experience is the best teacher: keep an eye out to see how much thin lines thicken, at what point very light lines are lost, or to what extent sharp corners or ragged lines in the original art become rounded and less ragged in the reproduction.

The use of materials broadly falls into two different categories: what will work and what works best. Large organizations often favor the use of less expensive materials that work well enough because of the high costs of volumes of material. Where the work is one-of-a-kind, the tendency is to use the best materials, because the cost is a fraction of the cost for the time that is invested in producing the product. Within reason, the same view holds for tools. When in doubt, get the tool that you know will work for you, because tools that are difficult to use slow your work and require more skill for a given level of quality.

The tools and materials described here are all useful and necessary for special purposes. You won't need many of them until you get to a specific situation that requires them. A simpler list might consist of the materials used for the project in chapter 1 plus a drawing board and a T-square. You can't hope to do professional-quality work without a T-square and a straightedge to work from. While a table can be set up to work as a straightedge, the best arrangement is to use a drawing board. The basic tools to prepare art for the printer's camera, therefore, are: T-square, drawing board, X-acto knife, and small ruling pen. The materials will vary with the paste-up you are doing, but basically they are: an adhesive, mounting board, drafting tape, nonreproducing blue pencil, kneaded rubber eraser, tissue paper, pencils, containers, rulers, and rubber-cement thinner. If the flat is porous and not too rigid, it is next to impossible to lift up art for repositioning without destroying the flat unless you use thinner. More specialized tools, discussed in this chapter, make your work quicker and easier.

Preparing Art for the Camera

Preparing art for reproduction on a printing press differs from preparing it for office duplicating equipment in that the art is first reproduced on film, which is then used to make the printing plate. The negatives can be manipulated in different ways before they are used to make the printing plate. This gives the printing press a definite advantage over office duplicating, in which a compromise in exposure and development must be reached in order to give the best overall reproduction. With intermediary negatives between the art and the plate exposure, the camera operator can shoot the art more than once with different exposures and developments and combine the best part of the negative from each shot for the optimum result. Each negative adds to the cost, but no compromise is necessary. Since commercial job printing involves larger quantities than office duplicating, the negative cost is not a large part of the total cost. Combining negatives is done for better quality, and the film itself gives finer fidelity of reproduction than duplicating materials.

ART NEGATIVE

3-2. A professional paste-up artist must be able to think in terms of the film negatives required to make the printing plate. This enables the artist to plan the art for the simplest execution.

The camera sees black and white, as with office duplicating procedures. Light grays and colors can fall out—that is, not reproduce. The reason for this is that the film used for print reproduction is a special kind that holds only a solid tone of black. There are no grays: the film comes out in the developer as either black or clear. This film is often identified by the trade name of the biggest manufacturer, Kodak, and is called Kodalith film. There are other manufacturers that produce the same kind of film, and all types have the same characteristics. This feature is required by the printing process, since the ink can either be printed or not printed—there are no positions in between.

Reproduction film is not panchromatic—that is, equally sensitive to all colors—but responds most strongly to red and least strongly to blue. Other colors fall in between. Colors also vary in density—the amount of black or white content—which affects the extent to which they reproduce on the film. To reproduce black lettering on a green field, for example, you should find out from the printer or camera specialist if the black lettering could be shot without the green background interfering. Color filters placed over the camera lens filter out the light from an unwanted color, leaving only the black image.

As with office duplicating equipment, for top quality it is best to work in black and white. The fact that printing reproduction film is not panchromatic is helpful in that it lets you use light blue for construction lines. The construction lines can be left in place because the light blue will not reproduce. On the other hand, construction lines for the printer or photo stripper are done in red so that they *will* come up on the negative. The red is a signal to the printer that these lines are for construction only: they are not to be left on the negative for reproduction. Red construction lines are used to indicate the location of artwork and for identity notations, trim lines, fold lines, and other information that the printer needs.

The camera operator generally shoots for the best reproduction of the main elements of the composition. For example, if the paste-up includes a mass of typography with very fine lines, the exposure and development time of the film will favor this, with other art left to come up as it will. There are many different ways in which the camerawork can be done, and not everyone sees it the same way.

If a paste-up includes art and copy, for example, the camera operator can shoot the art separately and combine the negatives. This is called *stripping* the negatives, and it is done during makeup and positioning for exposure onto the printing plate. The entire process is performed by the photo stripper. It is also necessary if the art is not going to be printed in the order in which it was prepared.

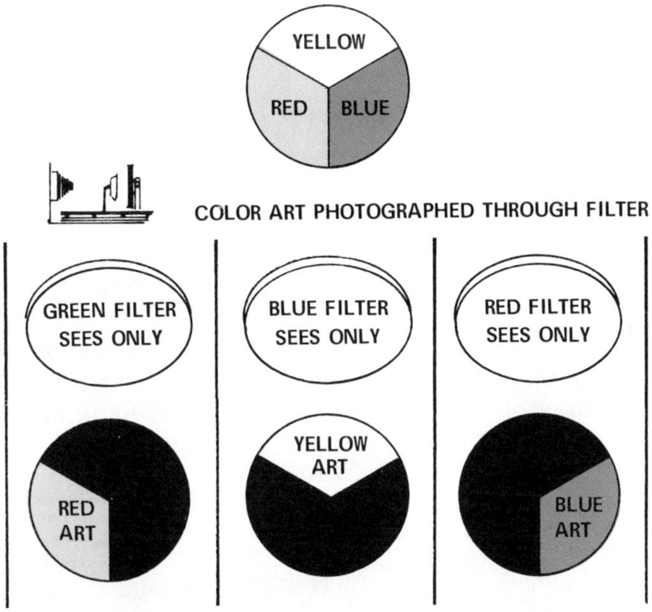

3-3. This diagram explains how each primary color is isolated so that it can be expressed in "black-and-white" film. The color filters are also used to separate colors out from art that is combined with a color.

Pages are usually pasted up in the order in which they will be read and seen, but the printer needs these pages in the order of the printing form for folding and binding. In other cases the printing may be done in multiples of two or more of the same image to fill the sheet without waste. Preparing and positioning the negatives before making the printing plate is called stripping and camerawork. If the paste-up artist has some idea of which stripping procedures will be used, it is helpful in preparing and presenting the art.

As a general guideline, the paste-up artist should prepare the art to use as few negatives as possible in order to save extra work, extra negatives, and stripping charges. Even if extra negatives are used for top quality, it is best to have everything exactly in position to guide the stripper and to let the camera operator determine whether extra negatives are needed and how to get the desired result. Some stripping departments prefer to prepare their own negatives at the same time for accuracy of register. They prefer to work to your guidelines on the master negative and paint up or remove panels themselves. This requires a master in which all the art is together on the paste-up flat. This technique is more expensive, because of the stripping work, but it guarantees the highest quality.

If you know that an extra negative is needed—if you have to reverse white type out of a panel and you are supplying only the positive and not the required negative image, for example—it is less expensive to separate the art into different shots. For instance, you might combine all the type that needs to be reversed into one shot, closely spaced to use a smaller negative, and let the stripper position it. Better yet, put it on an overlay in position if the size required is not excessive so that no extra stripping will be necessary. Again, as a guideline, prepare the art as completely as possible for the lowest cost.

The stripper will almost always have to do some retouching on the negatives to block out pinholes that occur during the developing of the film. The only exception to this rule is if the negative is purposely overdeveloped to thin the image. If the art is clean, there will be little to retouch, but dirt, shadow edges, lead-pencil construction lines, and the like can require a lot of retouching and result in higher costs. Keep your art clean!

The stripper can remove lines, panels, type, and other parts of the image. He can inscribe lines, change sizes, reverse, flop, reposition, and assemble the art. Further, the platemaker can remove elements from the offset printing plate even after it has been put on the press.

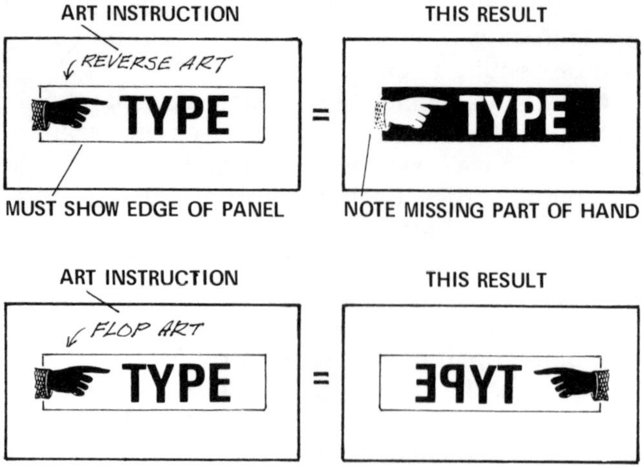

3-4. "Flop" and "reverse" mean entirely different things to the photo stripper. "Negative" is a piece of film.

Materials for Paste-up

The items used in preparing a paste-up, which are either incorporated into it or used up during the course of the work, are the *materials*, in contrast to the *tools*, such as a ruling pen, which, although they wear out eventually, have a more permanent nature. Ink, for example, is a material that is used up and is literally incorporated into the paste-up. This distinction is important for commercial work, because tax is not paid when purchasing materials that will be resold but is collected on the finished product. Tax is paid on the tools, because they are not resold. If you become involved in bookkeeping, you will have to keep track of this.

The materials used differ with the requirements of the individual project. In a commercial art department, for example, there is usually a policy as to which materials to use for a given assignment, and these materials are bought in bulk and supplied. If you do paste-up work on your own, you will have to decide for yourself which materials to use.

New art materials are always coming out on the market. There is a constant flow of new materials as different new techniques gain popularity. On occasion a material developed for one purpose is found to be excellent for a different, originally unintended purpose. An example of one such product is the Ulano Company's Rubylith and Amberlith. Both are special masking films developed for lithographic, engraving, and silk-screen stripping departments, and they can be hand-cut. They are "light-safe" stripping films coated on a polyester-backing sheet. A special adhesive allows stripped portions to be replaced on the polyester for corrections; the film can also be transferred to film negatives or positives. These products are a useful art material, whether or not this was the original intention, because the paste-up artist can cut lines and panels quickly and accurately without having to rule them and then fill them in. The material works nicely, is comfortably "see-through," and is on a stable base, which makes it very suitable for overlays. Any additional inking can be done on the frosted polyester base.

In the last several years there has been a considerable increase in the rate of development of photo and reproduction materials. The techniques used for paste-up have been affected and have as a consequence changed. This is interesting and helpful, but it means that the paste-up artist has to be constantly alert to these changes. It is probable that there will be continual experimentation for better or simpler materials.

The following is a general description of conventional, basic materials that you will encounter. They are the ones most often used, but there are many more special-purpose materials and a variety of

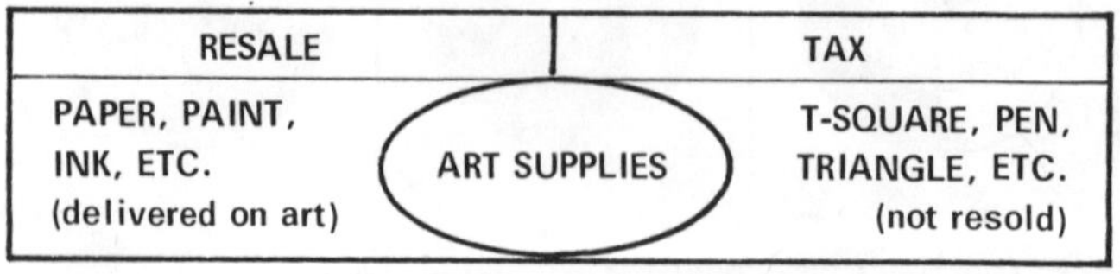

3-5. Important information for the accounting department (which could be you!). The government requires a different accounting for each category.

brands that have special qualities. A good-sized, fully stocked art-supply store has literally hundreds of items relating to paste-up. It is well worth your time to investigate different materials and the techniques used with them.

Always select materials that give a crisp, finished appearance to your work. Art is expensive and should look it. Paste-up is finished art, and while its main purpose is to provide the printer with the material needed to produce the finished product, it is also viewed as a piece of art—the final piece preparatory to the irrevocable step of printing. Good materials are expensive but well worth the price, which is minimal in proportion to the total cost of the job.

Adhesives—wax and rubber cement—are the two staples for paste-up. Wax adhesive is a more recent development. It makes a good bond for paste-up work when burnished down, is permanent, and does not discolor or bleed even through thin paper. It has the advantage of remaining tacky and allows the art to be shifted without leaving a residue on the mounting flat. To apply it, wax must be melted in a waxer and rolled on the back of the item that is to be mounted. Waxers come as small hand-held tools, such as the one made by Lectro-Stik shown in figure 3-6, or in larger sizes for commercial work.

Two types of rubber cement are available, regular or one-coat. Both can be applied with a brush; one-coat is stickier than regular rubber cement. One-coat rubber cement also comes in an aerosol-spray can. The simplest, least expensive adhesive is a small can of regular rubber cement with a brush applicator. Both types of rubber cement bleed through art after a couple of years and discolor. If it is left exposed, the adhesive tends to release in a shorter time, but if mounted art is stored in a folder away from light or air, this process is slowed considerably.

Both *single-edge* razor blades and *X-acto* knives and blades are popular paste-up materials. Some artists resharpen the blades with a stone, but most throw them away when dull—they are relatively inexpensive. You can break off the corner of a razor blade to expose a sharp corner, but be careful of your eyes and of others, because the blade can splinter. The #11 X-acto blade is the most popular,

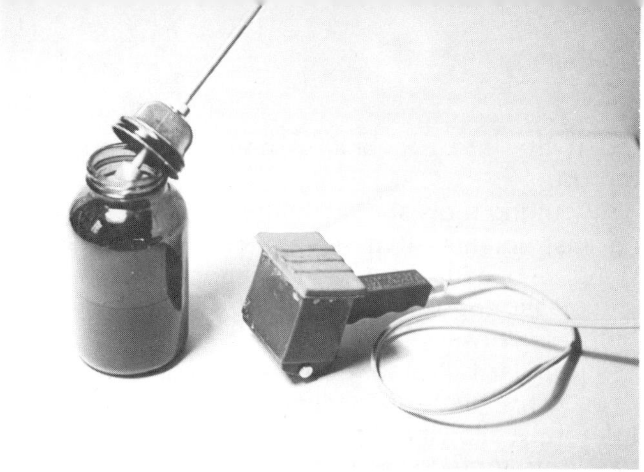

3-6. Hand waxer and rubber-cement jar are the tools for applying adhesive. Wax is preferable for volume work. This waxer is an antique once made by the Lectro-Stik Company.

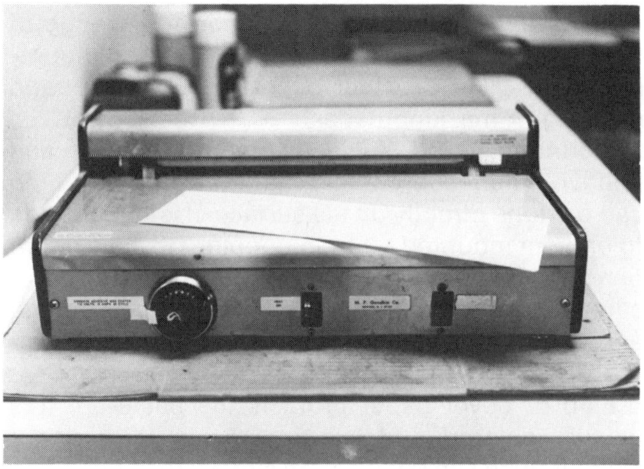

3-7. This commercial waxer coats the underside of galley sheets. It is used by typographers, but most artists use the slower hand waxer.

3-8. A fine-grained sharpening stone is necessary to hone a razor edge. Some artists use it to keep the X-acto knife sharp, but most just replace the blade.

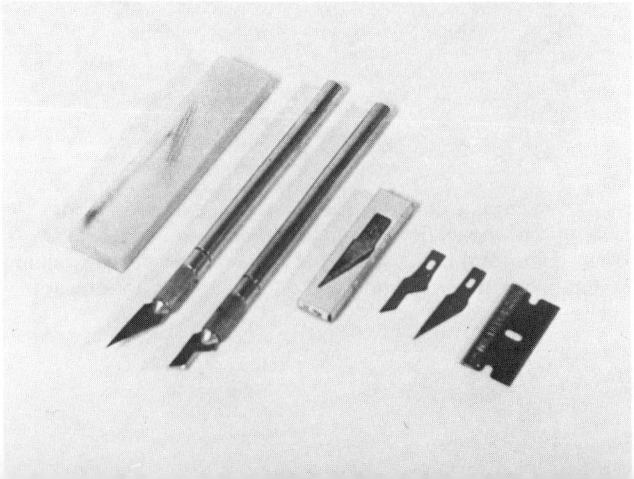

and it fits into a pencillike holder, allowing fine control.

All cutting tools should be used with metal drafting instruments—that is, a metal T-square and triangle—because it is next to impossible to work with plastic edges without nicking them, which ruins them for drawing. The exception to this, which is less popular but more convenient, is a #16 X-acto blade. This blade is smaller, shorter, and angled differently, allowing the heel of the blade to ride on a plastic straightedge so that the possibility of nicks is reduced. The ability to see through the edge of plastic tools makes the work easier. Metal drafting tools last a lifetime, but plastic tools are easier to use and far less expensive.

Art paper is available in an enormous range of surfaces, finishes, and colors. You will need mounting board, flap or cover paper, tissue paper, and drawing or masking paper.

Mounting board is usually smooth in finish and relatively nonabsorbent so that inkwork and ruling can be done directly on the surface. It is stiff enough to prevent mounted art from popping off if the board bends. It is nice to work on clay-coated stock, because inkwork can be scraped away for retouching and corrections instead of being whited out. Scraping is faster and cleaner.

Flap or cover paper protects the paste-up. It is usually colored and textured. Art studios use a specific type somewhat like a signature to create an identity for their work.

Tissue paper comes in two grades: tracing paper, which is thin and inexpensive, and a heavy vellum, which is used for overlays. Tracing vellum should not be used for overlays with a critical register, especially in larger sizes, because it shrinks and expands with moisture and throws off the register. For this kind of overlay you should use acetate.

Drawing or masking paper is popularly a heavy bond with a smooth finish, suitable for inkwork. It is used to block out portions of the art or for work that will be mounted in the paste-up.

Black *ink* comes in two varieties: waterproof and water-soluble. Red and nonreproducing blue inks are usually used in ballpoint pens, because they are used for construction lines, not for reproduction.

Waterproof ink is always best for reproduction, but it shouldn't be used in fountain-type drafting pens. Fountain-reservoir drafting pens require a soluble ink to prevent clogging, and the finer points usually have to be soaked in a cleaner from time to time if ink is left in the pen in order to keep them free-flowing. Shake the bottle of ink before using to maintain a dense black suitable for reproduction. Waterproof ink is used in a ruling pen that can be wiped clean, and it comes in very handy for difficult surfaces like acetate overlays.

Retouch white comes in jars or tubes. The tube variety is more concentrated, so you can mix a heavier consistency for quicker coverage.

3-9. Art paper is a complex subject: the six basic kinds for paste-up are shown here. From top to bottom: flap or cover paper, good (rag) bond, tracing or tissue paper, heavier vellum tracing paper, thin mounting board for flats, and hot-press illustration board.

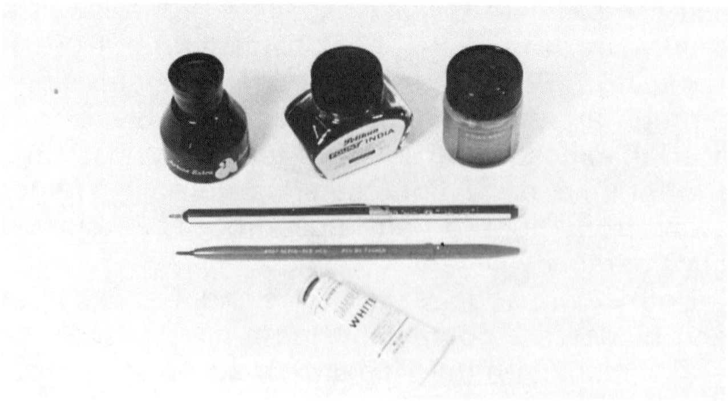

3-10. Some artists keep drafting pens filled with water so that they can use the denser waterproof inks. Unless you are willing to refill the drafting pen each time you use it to keep it from clogging, different kinds of inks are necessary.

The most important *pencil* to have is a nonreproducing-blue wax pencil. This is the most frequently used tool for drawing construction lines. Some artists prefer to use a 2H pencil lightly, because the lines can be completely erased for presentation of the paste-up. The softer HB-grade pencil is also popular. It gives nearly the same density of blackness as a #2 writing pencil but is noticeably better in quality. A B pencil is softer still, and as the number preceding the B becomes larger, the pencil becomes softer—6B is very soft and black and also hard to erase.

If your work requires marking up photographs, you will need a grease pencil. Do not use a marker that will mar the surface of the photo or that cannot be washed off with rubber-cement solvent for later remarking.

The most useful *eraser* is a kneaded eraser, which is used for cleaning and erasing. Don't use a kneaded eraser like an ordinary eraser—it really has to be kneaded. Hold it in your hand to make it warm, shape it into the most convenient form, then knead it like bread dough to produce a clean surface. A harder eraser, like the one on the end of a common pencil, will remove the more stubborn lines without damaging the surface of the paper, while an ink eraser is harder still and will remove the surface.

The most popular *pen* for drawing red construction lines is a fine-line ballpoint pen. Markers are handy for drawing art instructions on the tissue overlay. (It is helpful to know that permanent oil-based markers are soluble with turpentine.) There is a great assortment of pens and pencils and related equipment to choose from. Select and use the ones that work best for you.

Drafting *tape* is basic and is used to hold the art on the drawing board and to mount overlays and protective flaps. It comes in a less expensive buff color and in a smooth white finish that can be used to make corrections. You can draw ink lines on the white tape.

Overlay materials are frosty or clear acetate, the Ulano Rubylith mentioned earlier, and tracing paper. The heavier weights of acetate, measuring about 0.005 inch, are easier to work with, as is the heavier vellum tracing paper. Frosty acetate takes inkwork more easily than the clear variety.

If any of your artwork was produced with carbon, charcoal, graphite, or pastel, it must be fixed before handling to avoid smudging. *Fixative* comes in an aerosol can and is sprayed over the surface. There are two basic kinds of fixative: workover and plastic. Plastic fixative is used with strike-on proofs produced with carbon ribbon and, applied lightly, helps prevent smearing.

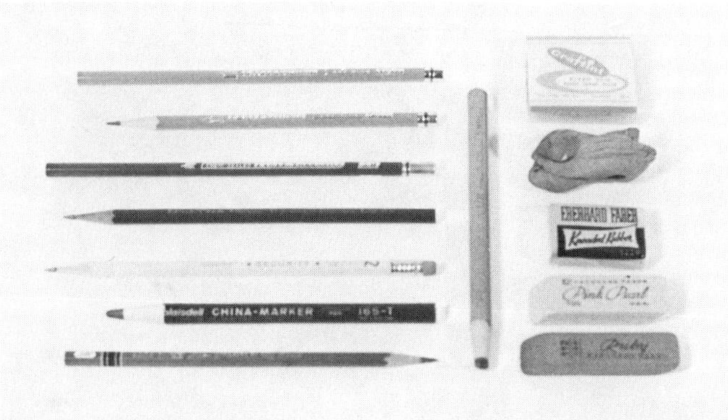

3-11. Pencils, markers, and erasers. Note the kneaded rubber eraser second from the top in the right-hand column; just above is the original package.

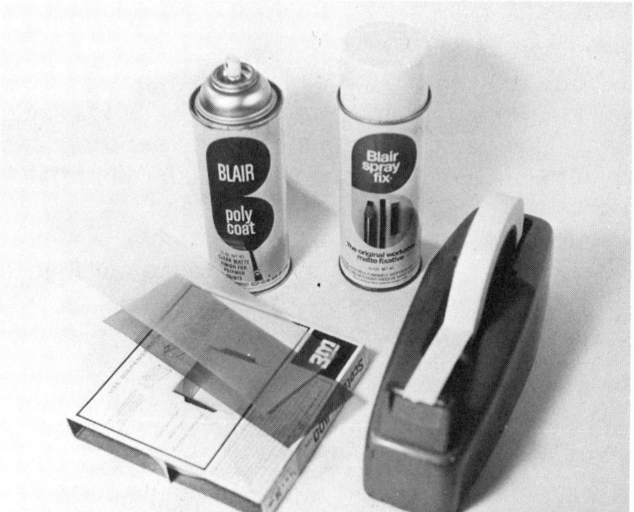

3-12. Tape and fixative. A tape dispenser allows you to hold the art with one hand while you pull and cut a length of tape with the other.

Tools for Paste-up

Tools are the more permanent part of the working arsenal. Some will last a working lifetime. A tool like the metal T-square or triangle is practically indestructible.

The working artist is constantly collecting tools that will make his work easier. He or she is looking for tools designed for a particular job or to solve a special need. As techniques change, so do the tools. Commercial work demands speed along with skill, and tools that facilitate your work are worth many times the initial investment. If you have the basic skills, you can do the job whatever method you use, but experienced artists tend to develop their own favorite methods to achieve a given result. For example, some artists correct a ragged black edge by running a brush with white paint along a ruler to make a straight line, while others rule the line with a ruling pen; some use a covering of white paper to make the edge, others use white drafting tape, and still others treat the edge as a basic correction, remake the art, and place it over the original. The same correction could also be made by scraping or erasing. There are seven different techniques to accomplish the same basic purpose. The results of each look different, but the appearance to the camera is nearly the same—the ragged edge has been corrected.

You can see from this example that there is room for interpretation, and there are reasons for using each technique. Each requires different tools and materials. Often a company or organization has a policy on preferred techniques, and as a consequence only certain tools are stocked. A professional paste-up artist should be accomplished in most techniques, selecting those that give the best results but able to adapt to new tools and methods that come along.

With the exception of the ubiquitous T-square and drawing board, the tools you choose are a matter of preference. Even such a basic procedure as using a razor blade to trim art materials can be discarded in favor of a pair of scissors. It is difficult to argue that a razor blade or X-acto knife is the only tool for trimming art (not that you won't hear this), even if it is the majority preference—some artists use scissors very efficiently. The tools described below are generally the most popular but by no means the only possibilities.

The *drawing board* and *lighting* constitute the working environment for the paste-up artist. Jobs in which these tools are substandard shouldn't be considered for permanent employment, because the psychological environment will be depressing. If you are providing your own tools, this is a good place to spend.

A good drawing board is sturdy and does not vibrate, holds its slant without slipping or loosening, and provides a true surface with square corners and edges. There are many different styles available in all price ranges. Lighting completes the environment—it is somewhat dependent upon many other external factors. To ensure adequate lighting, it is best to use a light fixture.

A number of drawing-board lights are available. They are designed to be mounted to the top edge of the board and are the most convenient lighting fixture, because they can be adjusted to keep reflected glare off the board and the work. Glare will quickly tire your eyes, so too much light is as bad as too little. Do not work with less than a 60-watt light 14 inches above your work.

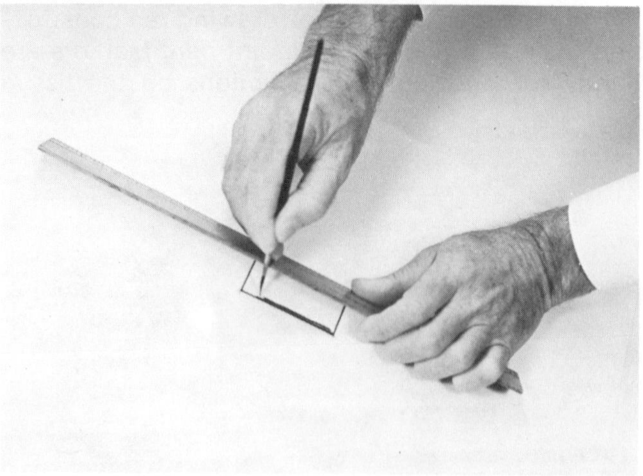

3-13. This correction technique is also useful for other purposes, such as filling a reverse panel. Brace the straightedge and guide the ferrule of the brush along the edge. The depth of touch is maintained by sliding your fingers along the top surface of the straightedge.

Another aspect of lighting worth noting is the effect of color. Continuous work with either the predominant yellow of an incandescent bulb or the predominant blue of a fluorescent light is not good. Use a daylight bulb with a better color spectrum or combine incandescent and fluorescent bulbs. Remember that fluorescent lights flicker—this can be very disturbing during sustained close work. Don't work in sunlight—if you try it, you'll find out why. Ideal lighting is daytime northern light, very nice if you have it.

You should plan on having this equipment, but if for reasons of cost you can't get it immediately, a tilt-up board with rubber feet that sits on a table is a good compromise. Use a comfortable three-way reading lamp on your left side if you are right-handed; on your right side, if you are left-handed. Arrange the chair height so that you are not hunched over and have a straight-on view of the work on the board.

A useful though not absolutely essential addition to the working environment is a taboret, which is a small storage cabinet placed alongside the drawing board. It provides both a storage area for tools and materials and a convenient working surface for tools currently in use.

A perfectly adequate plastic *T-square* can be obtained inexpensively. The qualities to look for are sufficient strength at the join of the two pieces to hold a position and a straightedge to use for ruling and cutting. If you sight down the edge like a rifle barrel, you can see how straight it is. A straightedge that is slightly raised from the drawing surface is better for ruling ink lines. If you use a reservoir ruling pen, make sure that the edge is not too thick for the length of the needle on the pen. Metal T-squares are more expensive. Lighter aluminum T-squares are not as impervious to nicks as heavier steel ones but are easier to use.

A *triangle* is used against the edge of the T-square as a guide for making vertical lines. By sliding the T-square up and down and the triangle right and left, all positions on the drawing board are covered. The triangle is also useful as a straightedge for drawing. Triangles come in two configurations, 30-60-90 degrees and 45-45-90 degrees. The most convenient size for paste-up is a 12-inch 30-60-90-degree triangle, because it covers the 8½-×-11-inch format that so much work is sized to. Only occasionally is there a call for something larger, such as newspaper-size vertical rules, and they can be drawn easily by turning your work sideways and using the T-square.

3-14. Convenient lighting is really worth the expense. The lip at the bottom edge of the board is an extra that catches runaway pencils.

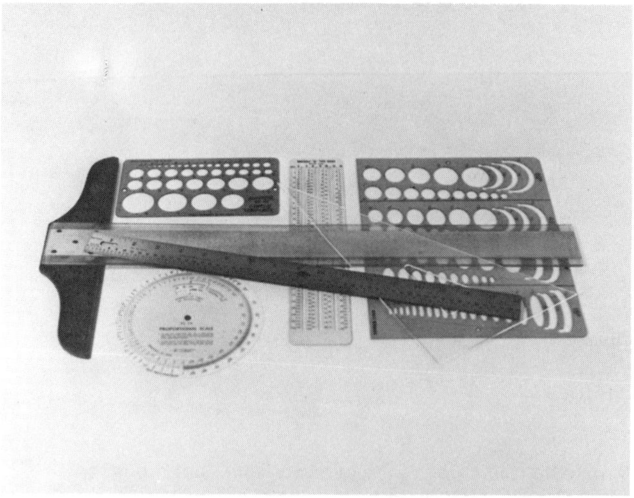

3-15. All of these tools are used quite a bit, but the real essentials are T-square and triangle.

Some artists use the T-square at the top of the drawing board to make vertical lines, but this is not recommended, because the board itself may be out-of-square. Using the triangle against the T-square assures a true 90-degree angle. The angle can be checked by drawing a vertical line and flopping the triangle over so that the bottom surface becomes the top surface: the vertical should be in alignment with the drawn line. A new triangle should be checked for accuracy this way. Of course, the T-square itself must be straight in order to do this!

As was mentioned earlier, triangles and T-squares can be made of either metal or plastic, and the lighter plastic tools cannot be used for cutting unless you have a small blade that does not touch the guiding edge. Some artists get around this by keeping two sets of straightedges, one for cutting and one for drawing. This makes sense, especially when you realize that durable metal instruments are ten times as expensive as lighter, more convenient plastic drafting tools.

You definitely need a ruler with a *pica scale.* A metal scale is very convenient because it can serve as a cutting edge when using plastic drafting tools. More important, the measurements are finely marked, and if they are indented into the metal, they will not wear away. Also the ruler itself is not subject to shrinkage or distortion. Measurements should be precisely in the center of the markings. If you are a little sloppy with successive measurements—say, the width of three columns of type and the intervening spaces—the accumulated error can throw the overall measurement off as much as 1/16 inch.

Scales are available in many forms and arrangements for special purposes. Two of the more convenient ones are the Haberule, used to measure line count and space for a column of type, and a proportion scale, used to find the percentage of reduction or enlargement in sizing art.

Templates, such as circle or ellipse guides, can be real time-savers for inkwork. A helpful trick is to put narrow strips of drafting tape on the bottom surface of the template near the guiding edge. This prevents the ink from bleeding under the edge if the ruling pen is not positioned at an angle to the edge—which is almost impossible to maintain for the full 360 degrees of a small ellipse, for instance.

You need a *ruling pen* for inkwork and a *brush* for white-paint corrections. Ruling pens can be bought separately—a complete set of drafting instruments, including ruling pens, bow compass,

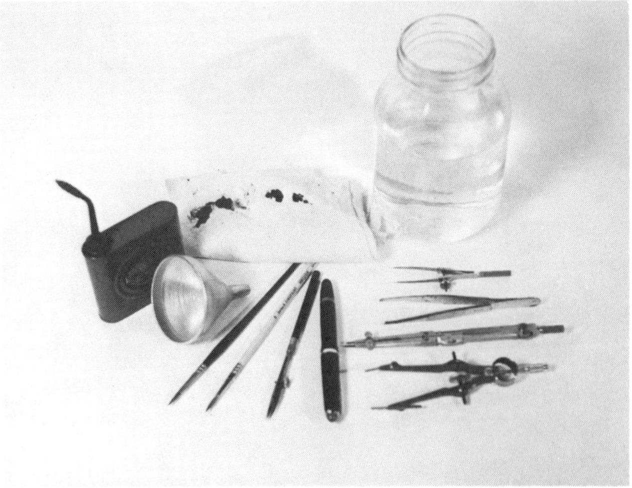

3-16. The ruling pen is the most versatile, but the reservoir drafting pen is preferred by many artists because it maintains a uniform line thickness and does not have to be refilled for each use. The circle cutter (bottom right) has a pencil-lead-size cutting blade.

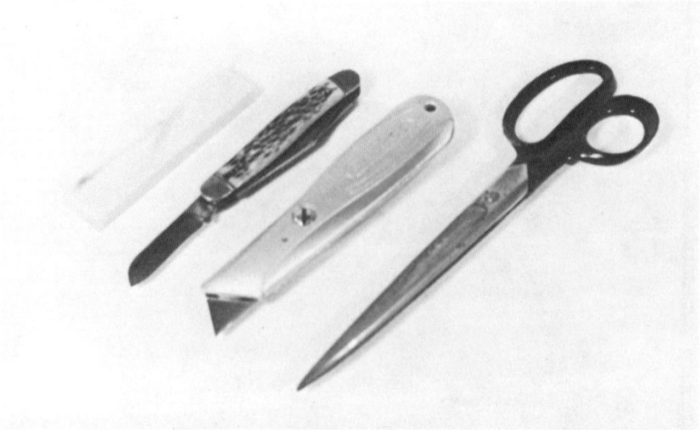

3-17. Don't be fooled by the pocket knife—it is a high-quality steel carving knife, and the blade is kept razor sharp with the sharpening stone.

compass and extensions, and dividers, can be quite expensive. Some artists wouldn't work without dividers, but a ruler can be used for the same purpose. Some work does require a compass, but a circle guide is often more efficient, particularly for small circles.

Most important is a ruling pen, which can be used on all surfaces and with paint as well as ink. Add a fountain-reservoir ruling pen in the #0 size, which is the smallest that can easily be kept free-flowing. Spend your money on a good white-paint-correction brush: get a #3 long-pointed sable brush. Nothing can be worse than struggling to make delicate corrections on inkwork with a soggy brush that won't point or hold the right amount of paint. Corrections can be exceedingly fine—you may have to clean up the inside of the letter "e" in ordinary text such as you are now reading, for example.

Use this brush for white paint only, not for black ink, because black residue will gray the white and make clean corrections impossible. An equally good brush is required for work with black ink. If you wash your brushes consistently after each use, they will hold up for many years of service. Of course,

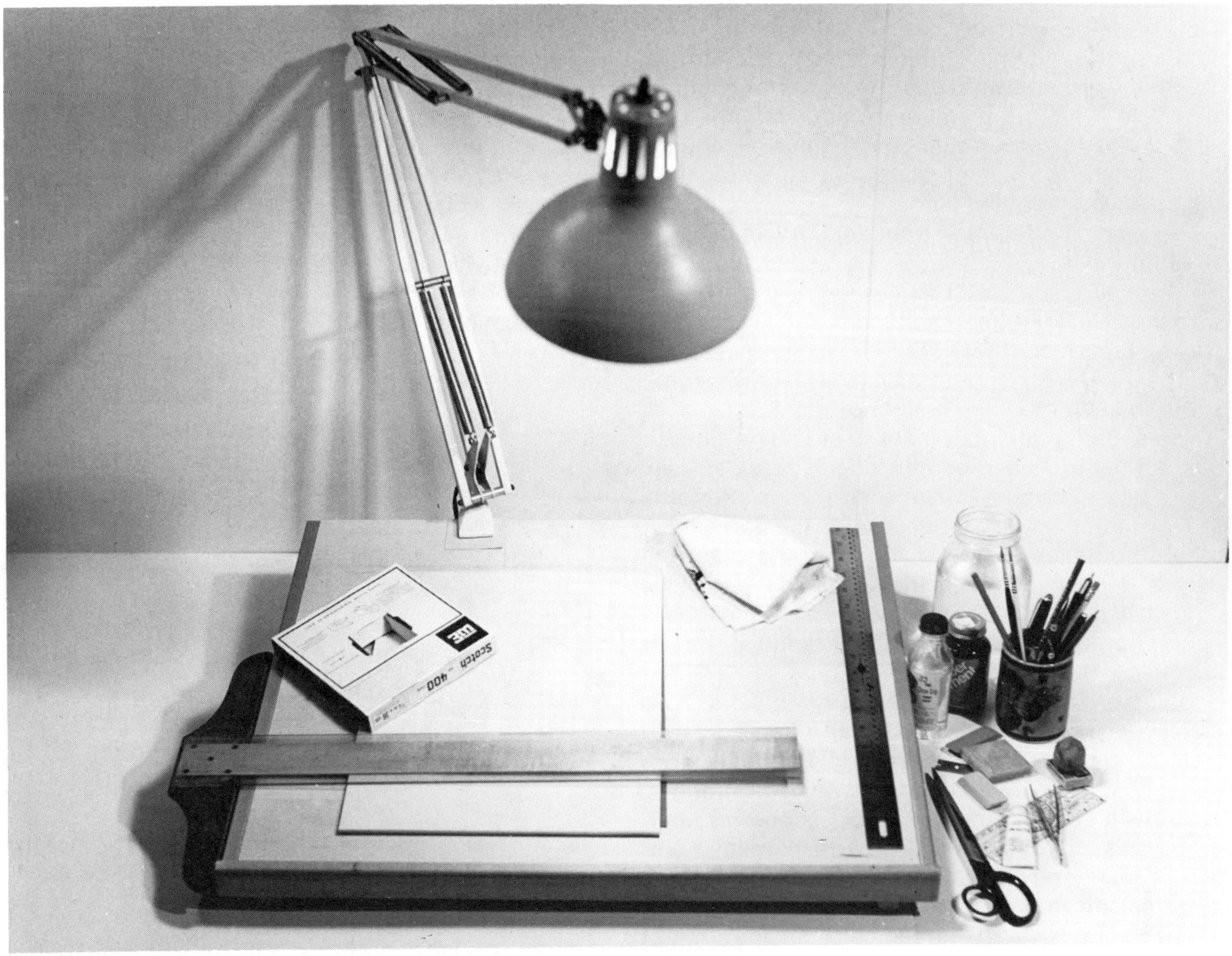

3-18. This equipment doesn't look as formidable as the taboret shown in figure 3-1. It is the minimum for professional-quality work. Be careful when you pick up a tool if you store pointed knives and pencils in a jar.

you should also acquire a jar for water and paint rags.

An *oilcan dispenser* for rubber-cement thinner and a *funnel* to fill it with are almost a must for making paste-up changes. The best ones for the job have a twist-shut valve at the spout to prevent the thinner from evaporating and to limit the flow to just what is needed.

It is helpful to use *tweezers* to handle art materials backed with adhesive. An old ruling pen with the set screw removed is excellent for this purpose.

A *circle cutter* is another useful accessory. It can be made from a compass by fashioning a blade on the end of a steel rod the size of a pencil lead, which is then used in place of a pencil. Some art stores carry these blades. Make sure that the compass will hold a radius when cutting pressure is applied.

Other instruments will come in handy from time to time, and there are many more than can be listed here. If you run into a situation with unusual requirements, check to see if there is a specialized tool that will save you time and labor.

The X-acto knife already mentioned, with a supply of different-sized blades, is very useful. You also need a sharp *mat knife* to cut larger materials to working size. Materials can be cut with shears on a cutting board, but these are more expensive than a sharp knife. You should also have a sharpening stone to keep the knife really honed.

You should remember that knives can be dangerous if they are handled improperly or carelessly. The easiest mistake, strange as it may sound, is to leave your fingers in the line of the cut. Check before you start to cut: hurried work is often the culprit. Second, do not let the tool get out of control or use it beyond its capacity to cut. Third, as a general practice, place the straightedge, guide, or safety on the piece you are cutting out so that bobbles or false cuts do not ruin your work. Two clean cuts produce better work than a single cut with irregularities due to loss of control. For better work and for safety do not cut beyond your own strength.

You will certainly need *scissors* sooner or later. Good scissors or shears can be resharpened and will probably never have to be replaced. A 6-inch blade is the most convenient size. Keep scissors sharp so that large sheets can be cut in one straight stroke, not chopped and chewed through, which produces a ragged cut.

Putting It All Together

Using the paste-up tools is the only way to develop skill with them. As you gain familiarity and confidence, they do not seem as difficult to use as they do at first. Rather than practicing with textbook exercises, which are isolated and irrelevant, use the tools to do a small newspaper ad. Your approach will be the same as a professional would use. Using the tools and techniques that follow, aim for professional-looking results. The design and art materials are given at the end of the chapter.

The drawing materials you will need are very heavy, smooth-finish drawing paper and tracing tissue. Use a board such as Strathmore three-ply kid-finish for the flat or mounting board; the thin tracing paper can be used for the protective covering flap. You will also need drafting tape, ink and retouch white, and the basic tools previously mentioned. You may want to take a few practice strokes with the tools and instruments before actually doing the paste-up, but work to the same specifications.

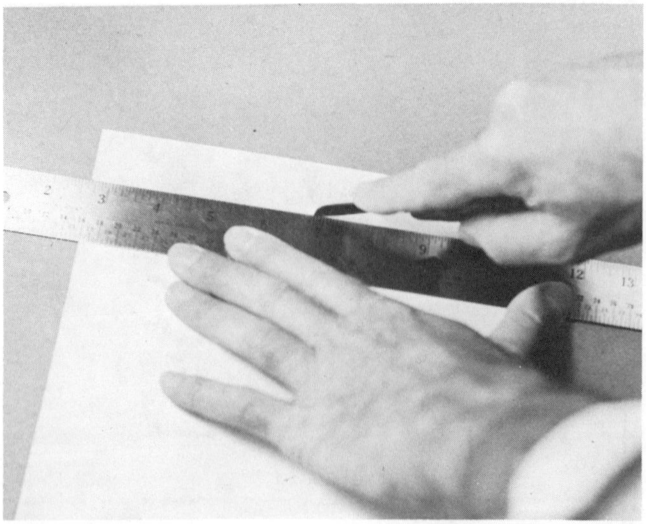

3-19. Hold the ruler firmly, keeping enough distance to move your cutting arm freely. Working on the floor is sometimes easiest.

The first step is to trim the flat material to the appropriate size for the design; choose something between 8½ × 11 inches and 5 × 7 inches. Using an illustration board or the back of the tracing pad as a cutting board, cut the mounting flat with the straightedge of the ruler and a sharp knife. Hold the ruler firmly in place and pull the knife toward you along the side of the straightedge. Use only as much force as you can safely control. Do not pull the knife toward the ruler: use the ruler only as a guide. Practice by cutting a strip from the edge of the board to get the feel. Cut to the length of the ruler. If the ruler is not long enough to go across the sheet, hold the knife in the cut and slide the ruler to a new position, leaving enough overlap to align with the previous cut, and cut again. Two or three cuts in a single position should go through the material. The reason for using a knife instead of a razor is that the handle provides a grip so that more force can be used and better control maintained. Again, do not use so much force that you can slip and cut yourself—hold the ruler firmly in place and be sure that your fingers are clear of the straightedge cutting guide. You can either measure and mark the line or use the T-square on the flat material to square the cutting guideline.

With the flat cut squarely to size, mount it to the drawing board using a strip of drafting tape at each side. Use a T-square to align the flat so that your work will be square to the edges. The following directions are keyed to the illustration numbers to show and tell you what to do.

1. Using the layout provided with the art materials at the end of this chapter, draw the guidelines and boundaries needed for positioning the art in nonreproducing blue. Measure the layout and use this measurement for the layout of your paste-up flat. Some artists use dividers when the art is this small. Some of the type measurements may be in picas, not in inches. To ensure accuracy, the flat must be drawn with the T-square and triangle. With practice this becomes second nature. If you are right-handed, hold the T-square in your left hand on the left side of the drawing board while you draw with your right hand. Position the triangle with your right hand and hold it with your left-hand fingertips—hold the T-square with your palm. Make sure that the T-square is always in full contact with the edge of the drawing board for true horizontals; likewise, the base of the triangle must be in contact with the T-square for true verticals.

2. After the flat is accurately drawn, the next step is to ink in the rules and the boundary on the reverse-

1. Draw guidelines in nonreproducing blue to define the design and to position the elements. Draw beyond the corners to help in aligning the type.

2. In this example the rules are inked first. Notice how the fingers of the left hand draw the triangle toward the T-square and maintain a slight pull on the T-square to keep it true at the same time. The motion of the pen is up, away from the T-square.

type panel. Use the practice strips cut from the flat material to get the feel of the ruling pen. The ruling pen is very simple to use. Just set the width of the point to the width of the rule, load the pen with a couple of drops from the dropper in the top of the ink bottle, and draw the rule. On most plastic T-squares the edge is raised from the surface, so there is little chance of wet ink bleeding back under the edge. Test to see how close you can draw to the edge before the rule will bleed. Use the straightedge of the triangle for this because it touches the surface. The pen works best if it is held almost vertically with only a slight slant to the right or in the direction you are drawing. Try to maintain a constant speed and pressure as you draw the rule, otherwise it will vary in thickness or darkness. Most artists overdraw the line at the ends and white back to the exact measurement to get a clean cutoff. If you are drawing very thin lines and have enough control, the line can be stopped at the proper point, omitting this step. Now ink the flat, using the triangle for the vertical lines of the reverse-type panel. It is usually easiest to draw the vertical line on the left side of the triangle from bottom to top.

3. Cut out the art materials and mount in position in the paste-up, as for the picnic announcement in chapter 1. If you are using a waxer and wax adhesive, it is easier to wax the back of the art before you cut it out. Don't use too much wax or it will squeeze out at the edges with burnishing.

4. Cover the back of the art with slightly overlapping strokes, then place on a cutting board and trim. Notice that the art should butt against the reverse-type panel. It will have to be trimmed closely and retouched so that it butts the panel cleanly without leaving a thin white line between the two units. The other materials do not have to be cut so close. Mount the pieces on the paste-up.

5. When the paste-up is completed and the rules and art are retouched, mount a tissue flap at the top edge, covering the paste-up, to protect it. Mark on the tissue over the reverse-type panel instructions to the platemaker that the type in the panel is to reverse to white letters on a black panel. Since more than one person may be preparing the film, mark in red on the flat alongside the art, with an arrow pointing to the type panel but not touching it, *reverse type panel* so that the stripper who handles the negative will know that the panel is to reverse and is not simply a rule border around the type.

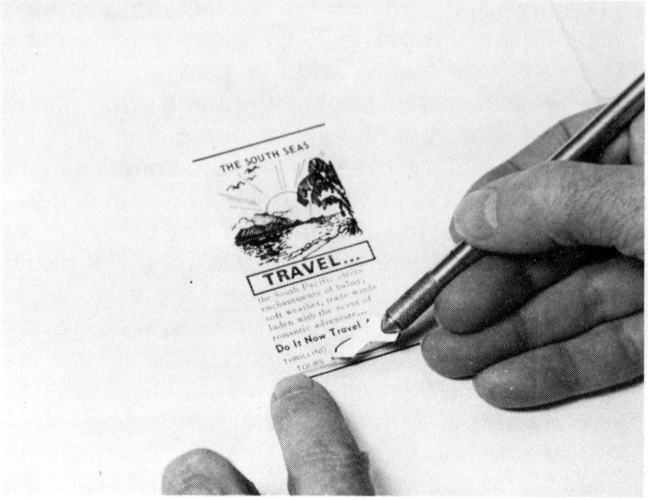

3. Place adhesive (wax or one-coat rubber cement) on the back of the type galley and picture and cut them out as in chapter 1. These elements are much smaller, however, and must be cut more closely and precisely. Tiny elements, such as the word being positioned in the photo, are best picked up using the blade as a shovel with wax adhesive or using tweezers with one-coat rubber-cement adhesive.

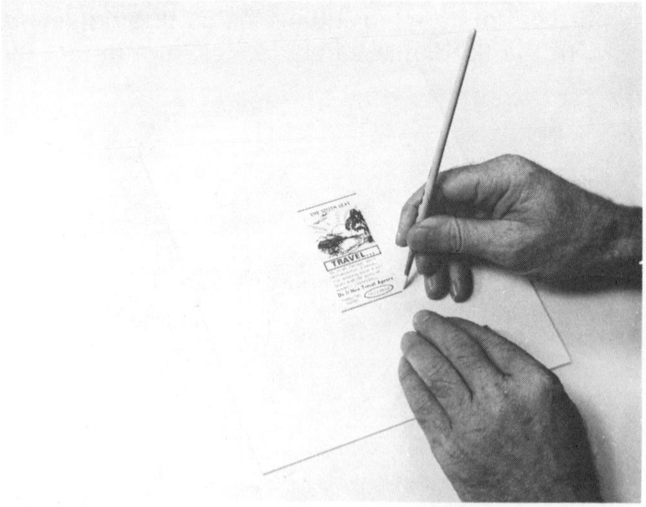

4. Cut the rules (extended at the ends) off cleanly with white paint and retouch anything else that needs it.

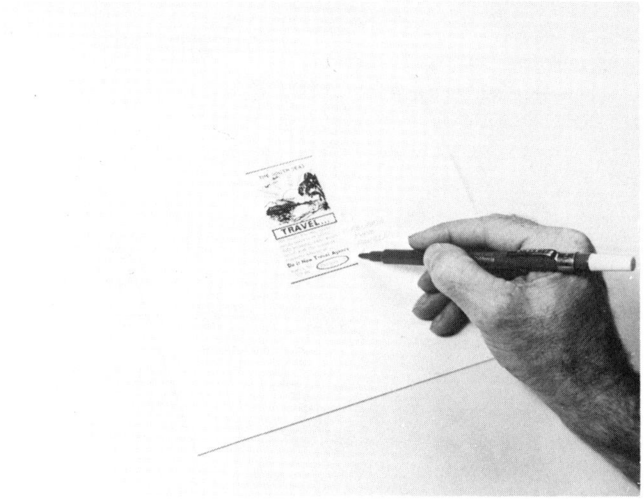

5. Letter instructions to the platemaker to reverse the type in the two panels. Draw arrows to the panels but do not extend them into the printing area.

6. After a final check, including the ad size, the paste-up is complete and ready for shooting. With the proper identification and publication information it could go to the newspaper.

No Mistrake Is Uncorrectable!

The only time it's too late to correct a mistake is when you find it in print! Not that this doesn't happen, but do try to catch them all before the paste-up goes to the platemaker. (Did you spot the typo in the above heading?) Once it is with the platemaker, mistakes are more costly to correct. The content of a paste-up is usually not the artist's responsibility, but sizing and graphic reproduction are.

6. Repeat the instructions on the protective tissue flap and make a final inspection to catch any errors.

The mistakes you make during the production of a paste-up are correctable, either by redoing the element and pasting it in over the error or by painting out the error with retouch white. If art is damaged during paste-up—if you cut something that shouldn't have been cut or scratch the reproduction image—it can be retouched with black. Naturally, such corrections are painstaking and time-consuming, so mistakes like these should be avoided, but it's nice to know that they can be corrected.

If a layout is the wrong size or a picture is wrongly cropped, the usable portion of the original art can be cut off the original flat and mounted as a unit on the new flat. Although all the image material should ideally be at the same height for perfect camera focus, reusing art will work if the flat is not thicker than three-ply. If the art is on illustration board, the backing must be peeled away for proper mounting thickness. The edges can then be sandpapered thin so that they will not cast a shadow line on the reproduction film.

Another correction method is to place a paper mask over the original art that allows only the correct portion to show. New material can be placed on top of the mask, thus remaking the art without disturbing what is underneath.

Practice increases skill and teaches you how to find the shortest route to the desired result. There are no formulas—any innovation that increases speed or improves quality is always desirable. Tools and materials are constantly improving and changing—use whatever works best for you. The purpose is what the art looks like in print.

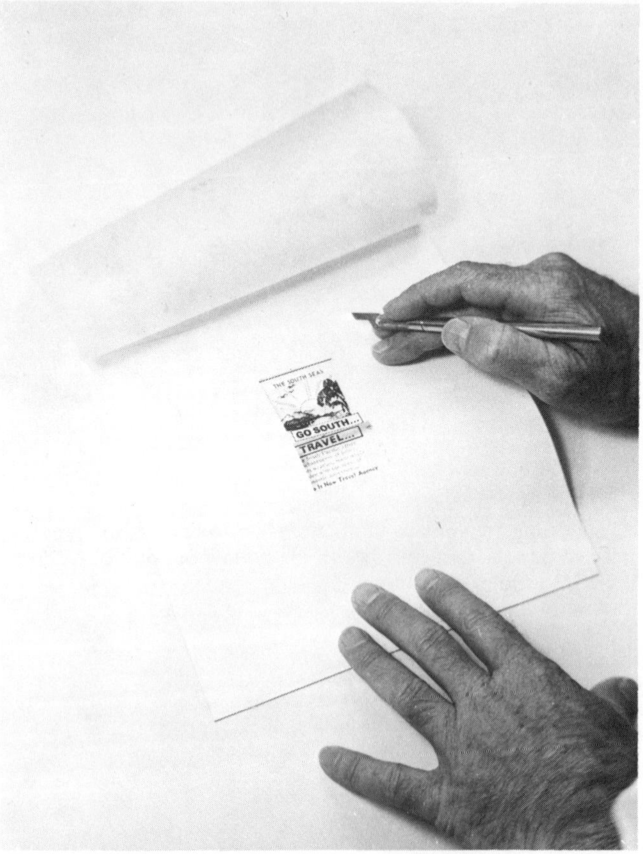

3-21. A paper mask with a heading change: when the mask is dropped into position, the new heading replaces the old, and the rest of the ad is the same.

3-22. Providing reverse photostats on the paste-up would make the art camera-ready: that is, no stripping would be necessary. Photostat reverses are suitable for instant printing. The term "camera-ready" is ambiguous, since any finished art is actually ready for shooting, and is best used only in relation to art for instant printing, with which no film manipulation is possible. There is a duplicate of this clip sheet at the end of the book.

CLIP AND USE THIS PAGE FOR MATERIAL TO DO PASTE-UP

LAYOUT

TRAVEL...

THE SOUTH SEAS

Do It Now Travel Agency

the South Pacific offers
enchantments of balmy,
soft weather, trade winds
laden with the scent of
romantic adventure...

THRILLING
TOURS

PAGO PAGO

TYPE GALLEY

ART

REVERSE STAT

PASTE-UP

THE SYSTEM
REV PANEL
5X7"
RULE ALL AROUND & BOTTOM

4. Basic Techniques

Paste-up as a full-time job requires professionalism. There is usually a considerable volume of work—with a publication, with catalog sheets, or with any printed matter that communicates technical information. A heavy volume of work requires a production system of some sort. A production system also has a tendency to simplify the work, so that fewer individual decisions have to be made. This helps speed up the work.

Speed and accuracy are the hallmarks of professionalism in paste-up. Watching the work of a professional paste-up artist is comparable to listening to music: you are aware of rhythm, harmony, flow—a sense of beauty. The difference, of course, is that your eyes are doing the "listening"—the aesthetic sense is expressed visually. Good paste-up work expresses the harmony of proportion and spacing that characterizes beautiful design.

The essentials of aesthetics and taste, a sense of beauty that is seen and felt, cannot really be taught. It is more accurately something within the individual that is, in this case, expressed visually. There are, however, techniques that aid your ability to express yourself. The procedures will help you direct your reasoning to the matter at hand.

There is little value in comparing your work with that of others, because the sense of beauty is unique for each individual. There can be technical comparisons—rates of speed, etc.—but these fail to touch the central point, the quality of the work, which is primarily a product of the aesthetic sense of the artist. Each artist expresses this quality in an individual way—a characteristic style.

It hardly matters how good a design is originally if all the subtleties are lost in the final reproduced product. The paste-up artist is the person who produces that final product, and his work is the work that is seen by the public, no matter how many previous decisions have been made or how many great ideas have previously been developed.

You may think that if the paste-up artist is all that important, how come they use a designer or other artists in the first place? This transition is exactly what is happening right now, and the paste-up artist's role is growing in importance with each change in graphic technology. In today's graphic industry the paste-up artist is often the only artist.

4-1. A professional paste-up artist often has to interpret rough layouts and arrange the design elements from written instructions alone.

Never let it be said, however, that this place in the sun is a personal burden. Professionalism implies, among other things, an ability to detach yourself from those forms of involvement that hinder your work and make it a burden rather than a job. The paste-up artist's responsibility is to maintain his integrity to his aesthetic sense, his sense of beauty, and to express it as best he can within the conditions required by the job at hand. Simple? Hardly.

Just one of the pitfalls is to imagine that your individual aesthetic sense is identical with the work that you are employed to do. If you are not in complete agreement with the ideas you have to work with, then this approach, to say the least, can dampen your enthusiasm. A professional need not find his identity in the work he or she produces. Your aesthetic sense is employed in producing the work—it is not the work itself. What is within the artist has a potential that extends beyond how it is expressed in a particular job. The door is just as open as you want it to be. The idea of noninvolvement becomes clearer as you go along.

Speed and accuracy, once you get the hang of the technicalities, develop naturally. To begin, forget about speed and concentrate on accuracy. Work that is clean and accurate is always usable, no matter how slowly it is executed, and preparing art for reproduction is the point of the whole thing. If you think about it, you can see that the more accurately your work is executed, the less reworking it will show and the cleaner it will be.

Professional work has a certain sparkle and crispness. It looks and feels somewhat like a brand-new book the first time you open it. It looks fresh and clean in spite of whatever problems have occurred during the course of its production. Maybe you've noticed that art pencils don't have erasers. This is not to say that professionals never make mistakes but that corrections are not simply smudgily erased. They must be so artful that they cannot be noticed. A certain amount of technical know-how can help you here.

Technical rules are made to be improved upon. Don't worry about making mistakes at the beginning—they are all correctable. Just aim at the center of the target, which is accuracy and good craftsmanship. They will develop with practice, and so will your speed. This is the key to professionalism, because volume work is the biggest area for the art of paste-up. Whether you are involved as a specialist or simply want to understand how the art works and have other goals in mind, paste-up is the fundamental for preparing art for reproduction in the graphic industries.

The biggest portion of paste-up is concerned with the communication of information, and the essence of this is the written word. By the time written material reaches the paste-up artist, it is in the form of *typography.*

Typography

There are three kinds of type proofs, produced by three different typesetting processes, that you may encounter. A type proof produced by the strike-on cold-type process will smudge: it should be coated lightly with a plastic fixative to protect the image. Always work with clean hands, and don't touch the typeset material. Type proofs produced by the hot-metal process must be thoroughly dry before you handle them—they smear very easily when wet. They cannot be sprayed with fixative, however, be-

4-2. Galleys are produced by three typesetting processes: cold-type photocomposition (left), strike-on cold type (center), and hot metal (right). Each is handled differently.

cause it will prevent the ink from drying, and once the ink is dry, the fixative is no longer necessary. Hot-metal proofs usually come in sets of three, so you have a little leeway. Proofs produced by the photosetting cold-type process are not subject to smearing, and the surface of the paper is hard, like a photograph, and easier to work with. You should be careful with the cutting, however, because, as with strike-on type proofs, there is usually only one set to work with.

Type proofs may or may not be backed with a wax adhesive. Having them waxed at the typesetter saves you the job of waxing each proof individually with a hand waxer. Whether or not proofs are prewaxed depends on the production system you are working with, and some artists prefer to use rubber cement.

To prepare proofs for paste-up, one helpful technique is to place the proofs face up on a board and align them to the board with a T-square. They should already be backed with adhesive, either wax or one-coat rubber cement, and ready for mounting. (Do not use regular rubber cement for this technique.) Make horizontal cuts for each block and column of type, holding the proof heavily enough against the board with the T-square so that it doesn't slip. Turn the board 90 degrees—at a right angle—so that you can keep cutting horizontally, and cut out the vertical portions of the proofs. If you are dealing with a lot of little blocks of type such as captions, cut them on one side only—say, the left side if they are positioned flush left—don't make the final cut until you are ready to paste them down so that you won't lose them.

With this technique you can trim all the proofs evenly and square to the type image. Although the camera should not see the edges of a type proof, you should keep them even anyway—edges that are just slightly crooked can throw you when you align the type. Another advantage of this technique is that you can easily and quickly pick up what you need, cleanly trimmed and ready to use. All the proofs are also in plain view so that you can easily keep track of the job—a very helpful point if you are working on a large project such as a publication.

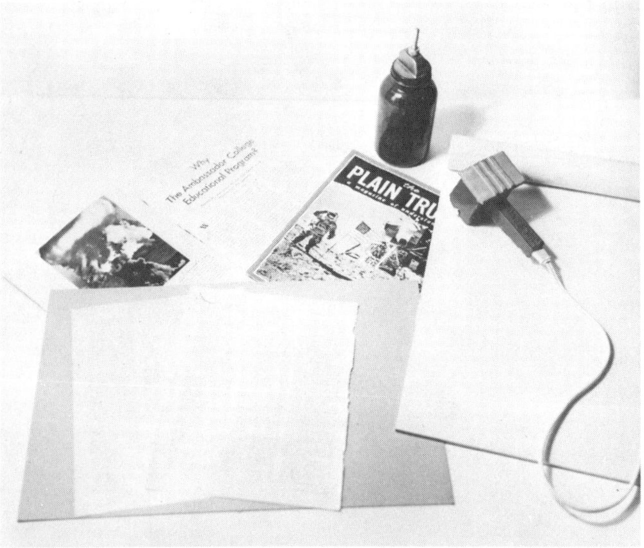

4-3. Volume material is usually wax-backed by the typographer. For this exercise prepare pages of type from a magazine printed on good paper.

4-4. Mass-production trimming is quickest and permits a visual check of the paste-up elements. For volume work, such as a newspaper, a mechanical proof trimmer is used, but advertising work, with its variety of type widths (measures), is handled this way.

Leave enough paper at the edge of the type image not to crowd yourself, but not so much that the edges of adjacent copy blocks overlap each other. The exact border width is determined by the particular layout design, but an average might run between 1/16 and 1/8 inch. If the copy blocks are positioned one pica apart with a rule in between, however, you'll have to trim the proofs closer than that.

To get the feel of this technique, try it out. It would be nice to use actual proofs, but they can get expensive. Instead, get hold of a periodical that uses good-quality heavy-coated paper—*Holiday* or *Vogue* magazine, for example—and pull out a few pages that have columns of copy. Remember, use good-quality *heavy* paper, not good-quality thin paper. Follow the instructions along with the keyed photographs.

1 and 2. Apply adhesive to the back of several sheets and place them on a board for cutting. The cutting board can be anything that you won't cut through and that you don't need for anything else. Set up a paste-up flat or board approximately 8½ × 11 inches in size. Using a nonreproducing blue pencil or ballpoint pen, draw in guidelines for positioning three equal copy blocks side by side. Work in any way that feels comfortable to you—there are no rules for this kind of thing. Repeat on another flat, this time placing the copy blocks one pica apart with a rule the depth of the block in between—a newspaper-type layout. Magazine paper is not as good as proof paper and for this reason is harder to work with. The image is also less sharp, because it is several times removed from the original. If the edge of the paper tears as you are cutting, the cutting edge of the blade is either too vertical or simply not sharp enough.

4-5.

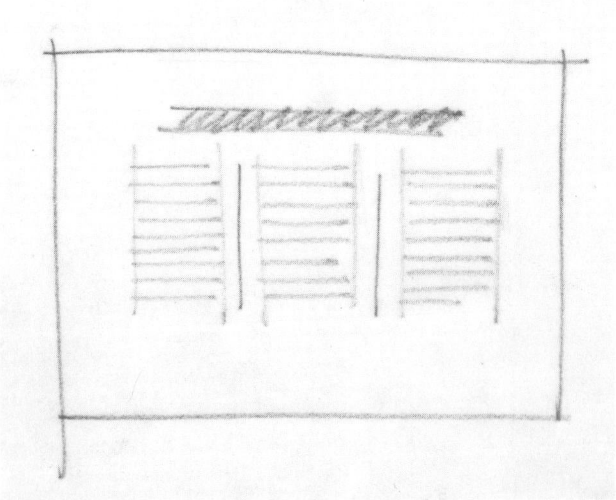

1. Using this thumbnail sketch as a guide, draw the paste-up flat as described in the text. Notice the number of decisions involved if the layout is not exact.

2. The process of drawing the paste-up flat is the same as in chapter 3. Each artist handles T-square and triangle a little differently.

3. The arrangement of the copy blocks usually looks best if the space between the blocks is slightly larger than the space between the lines of copy. The depth and appearance of the copy block are functions of the size, leading, and character of the particular typeface you are working with. There are many ways to divide a column of type into equal copy blocks. The most direct method is to count the lines—say, three hundred—and divide by the number of columns—say, three. Count off the result—one hundred—in the column of type three times. If you want ten lines per column, for example, count off again by ten: you will have thirty lines per page and ten pages. Or, if you want five pages, count off by five: you will have twenty lines per column. For a simple job you can mark off the column depth on a piece of paper and use it as a ruler—or you can use a ruler. Some artists prefer to use dividers or a Haberule—a rule that measures line count. Others hold the trimmed proof to the flat and walk the type through.

4. As you cut between the lines of copy at the proper depth, you will notice that the space between the lines varies with different typefaces. Some typefaces, set solid, provide as little as 1/100 inch for cutting space between the ascenders (d, f, h, k, l) of the lower line and the descenders (g, j, p, q, y) of the upper line, so cut accurately, or parts of the letters won't be there when the camera is ready to shoot them.

3. Of the many ways to determine line count this method is at least exact. But since lines and paragraph breaks seldom fall evenly, visualization is still required.

4. Try to cut between the lines at exactly the same depth. This makes it easier to move or respace them. The precision required is similar to the example in chapter 3.

5. Positioning the copy blocks takes practice, but an experienced artist often aligns them correctly the first time.

5 and 6. The next step is to align the copy blocks with the T-square. If the guidelines and the position of the copy blocks are accurate, you may not have to move them, but this takes practice and an ability known as "T-square eyes."

7. Using the T-square as a fulcrum and the blade of the knife or a pick as a lever, move the proof very slightly until the base of a line of type is in line with the T-square at the proper height and position. It takes practice to hold the T-square properly and to know how hard to press down on the proof so that it sticks and yet is still able to be moved to the final position. Some artists like to use tweezers for this part of the work; others use their fingers. Whichever way you work, avoid touching the type image with your fingers, because you cannot lift off the dirt without lifting off the type image as well. If you are using rubber cement, tweezers are a convenient tool.

6. If the copyfitting calculations are difficult to visualize, trial-and-error often will prove to be the quickest way.

7. With the type in position by eye, align it precisely with the T-square. Even experts check their work this way.

4-6. If wax-backed proofs were moved across an area to be inked, the ruling pen would still make a line. A drafting pen will skip over anything waxy or greasy. Test the line width before inking.

After a few attempts you will probably be able to drop the copy blocks in place easily. The next step is to position them closer and put in a rule. It is hard to ink a rule over grease or wax—if you use wax adhesive and are putting in rules as the final step, avoid sliding the wax-backed proofs over the mounting board. If you are using photoset proofs, however, you can wipe the wax residue away with a clean rag or tissue moistened with rubber-cement solvent. This is impractical with other kinds of proofs or with publication pages, because the solvent will dissolve the ink or the carbon and smear the type image. The best approach here is to draw the flat accurately and ink the rules in at the beginning. If this is too difficult or time-consuming, just draw the rules on a sheet of proof paper and mount them as if they were type proofs. Remember, too, that you can use this method to correct a mistake.

4-7.

1. Sometimes the rule can be drawn at the waste edge of a type galley. Cut it into segments for paste-up, which guarantees that each rule is exactly the same thickness.

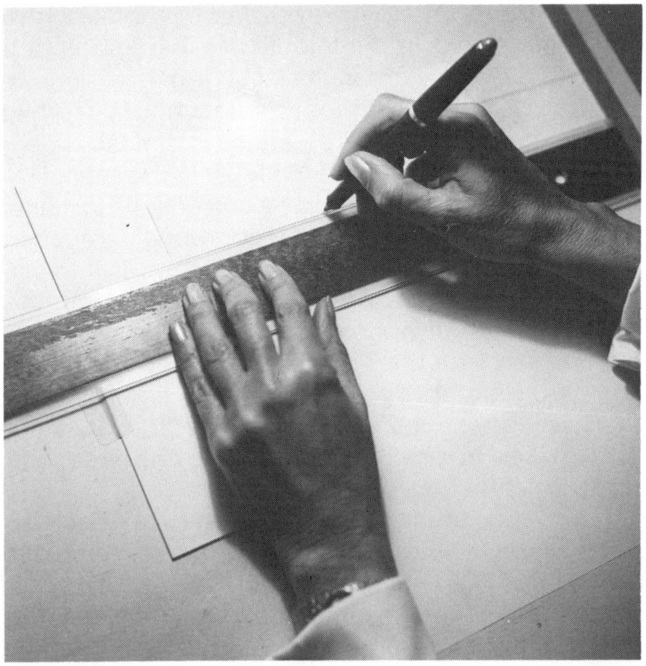

2. The problem with this method is to get the rule exactly straight. The advantages are uniform thickness and a clean, square cutoff at the ends.

To paste up a rule, draw it longer than you need, place it just above the right position, pull it down straight, and, maintaining a slight tension, drop it onto the board. Maintaining the tension ensures that the rule is straight and is much easier than trying to slide it into position. Trim the ends to the proper length.

To draw in a rule, the easiest method is to use a drafting pen such as a Rapidograph. If you have to draw a rule on a difficult surface or in a material such as paint that can't be used in a reservoir pen, use a ruling pen. The rest is just practice. One little hint—if you are going to rule on a finished paste-up, test a few lines on a piece of scratch paper.

You are now on your way to becoming a paste-up artist. The old adage "practice makes perfect" is the key. To be sure, some people seem to find it easier than others, but dexterity and skill aren't everything—visible aesthetic quality is what really counts. The artist is constantly making little decisions about the position and arrangement of body copy. Not the least of these is the necessity of having the copy read through.

4-8. Burnishing is required to make either wax or one-coat rubber cement bond to the flat. Use a protective sheet—one with a hard finish like tracing paper works well.

Paste-up artists usually don't check copy continuity, because their responsibility is for the graphic and not the literary content. Because an artist looks at copy as a visual element, it is easy to misplace copy blocks, so do read at least the last and first lines of the copy to make sure that it is in sequence. This should be done before the final *burnishing*—rubbing the proofs down firmly to bond the adhesive.

It would be nice if the artist could catch typos as they went across the drawing board, but in practical terms this is asking too much, and the normal practice in commercial situations is that the editor or writer reads the copy. Often several people will check the art. There are some errors? Why can't they be caught *before* paste-up?

Corrections

One problem with aligning columns of type, which you have probably found out, is that it is difficult to get an exact line count. You often have to move a line from the top of one column to the bottom of the next, and a single line of type is hard to handle—the best way is to position it as you would a rule drawn on a separate piece of paper.

Editors and writers have the same problem with a word misspelled in the middle of a paragraph or a widow—a single word in the last line of a paragraph. If this widow leaves an inordinate amount of white space, that has to be fixed too. It is almost impossible to plan for how the type falls, and it seems inevitable that there will be some corrections to make.

Typos corrected by the typographer are often done by resetting the line and pasting it into the correct position on top of the line containing the error. Be careful not to shift these lines or knock them askew with the T-square when you handle the galley during the paste-up. Some typographers mortise corrections into the galley and hold the new pieces in position by backing the area with tape, which helps considerably.

Without a doubt, the best way to make corrections is on a line-for-line basis by resetting the line. For one or two corrections, however, it often is easier, less time-consuming, and cheaper to make them at the paste-up board, especially if an extra paragraph of the galley has been set and the required material is available.

One neat way to make a patch correction, which ensures that the tiny letters won't get knocked off, is to cut the proof above and below the line to be corrected, just as you would to move a line of type except that you start the cut at the end of the first word and end it at the beginning of the last word (figure 4-9). Carefully make vertical cuts on each side of the offending letter and lift it out with the point of the knife. Cut the replacement letter at exactly the same points—say, just above the ascenders of each line and exactly at the side of the letter—pick it up at one side of the image with the point of the knife, and drop it into place. It should fit perfectly from top to bottom and be at the right height.

The spacing of a line can be adjusted by slitting between the words and changing the distance between them to equalize the space in the line. This process is easier with a horizontal slot to slip the words along. If you have to reduce the space between words, cut off a little segment of the proof between the words, just as if you were lifting out a letter. Burnish the proof, making sure that nothing slips out of position.

To get rid of a widow, the copy has to be edited, usually by eliminating words. You can handle individual words with the same technique that is used for individual letters. Often an entire clause is eliminated so that the sentence ends at a comma.

You can save yourself the task of dropping in a period if you are working with proofs on clay-coated stock by scraping away the tail of the comma—presto, a period! If the copy cannot be edited, at least the last line of the paragraph must be reset and placed over the original.

Eliminating a widow by editing leaves the copy column one line short. This means that the line has to be taken from the continuing column, and the space positioned however it looks best. If possible, leave the line space just above a heading.

To demount a galley from the board for corrections, lift up the corner and spread a small amount of rubber-cement solvent on the underside. As the solvent dissolves the adhesive, the galley will be released. Pick it up, let dry, and reposition. A word of caution—if the solvent gets on the type image, do not touch it until it dries, or it will smear (unless it is a photoset proof, which is impervious to the solvent). The easiest way to handle the solvent is with an oilcan.

To make corrections on body copy, where
not → the entire line is npt reset, this technique
can be helpful, ~~and offers the advantage of~~
~~preventing~~ the type from ~~getting knocked~~
~~off~~.

4-9. Two correction techniques are shown here: for the misspelling of "not" on the second line and for the editing change to remove the widow on the last line.

4-10. A few drops of rubber-cement solvent will release both wax and rubber-cement adhesive. Be careful not to mutilate or smear the type.

Headings and Spacing

Pasting up headings is a lot easier than pasting up the smaller body type, except that you have to decide how to space them. If you are working with a tight layout, there should be no problem. However, in many paste-up situations there may be no specified layout or it may be done very loosely, leaving the exact spatial relationships undetermined.

Flip through the magazines you were using for type proofs and look at the headings. Your eye always goes to the heading before the body type—not just because it is first but because it is positioned in such a way that it *seems* first.

Try this experiment to prove the point. Cut out a heading from a magazine and place it above the three columns of type you previously pasted up. Ask two other people to do the same thing, preferably with a heading from the same magazine but at least with a similar heading. Each of you should work independently. Using dividers, compare the spacing of the headings. You will probably be surprised at the variation. The greater the skill of the artists, the more their work will be in agreement. This may seem surprising, but it is reasonable from the standpoint that the more trained eye, the more it senses "right" spacing.

There is no such thing as "correct" spacing: it is a matter of fitness, an exercise of aesthetic judgment. There are no real rules, but perhaps a few generalities will provide useful guidelines. The spacing within the heading should be greater than the spacing within the body copy. Since the heading is usually in larger type, it will require more space between lines than there is between the lines of body type, and, again, there should be more space between the heading and the body type than between the lines of the heading.

Subheadings should relate more to main headings than to the body type, which means that there should be slightly more space between subheads and body copy than between subheads and heads.

4-11. Try the experiment in spacing a heading with several people and compare the results. It reveals design sense and something about personality too.

4-12. The usual arrangement of type and general rules for spacing are similar to the Ayer format (figure 2-8).

Don't take these generalizations as hard and fast rules—you can probably find an exception in the magazine you used for your experiment.

One spacing requirement that is more definite is centering a heading. Center means in the center. As with determining line counts, there are many ways to do this, but one of the easiest is to draw a centerline in nonreproducing blue and then position each line of the heading equidistantly to the left and right of the centerline. Look out for certain combinations of letters that don't look centered even though they are, especially lines ending or beginning with punctuation marks. Adjust the literal center as necessary to get the right visual effect.

If these techniques seem a little vague or inexact to you, remember that paste-up is artwork. Pasting up a page of type is really an exercise in art composition. The freedom of art and design applies equally to working with type and to paste-up. The only difference with paste-up art as compared to drawing or painting is that the final product is once removed. The paste-up is done for the camera and the platemaker, and the final product comes off the printing press. For this reason, the art is handled in a different manner than it would be if the paste-up were literally the finished product.

4-14. Centering can be done with a ruler as well as with dividers. A centering ruler starts the measurement from the center of the scale and gains units equally to the right and left.

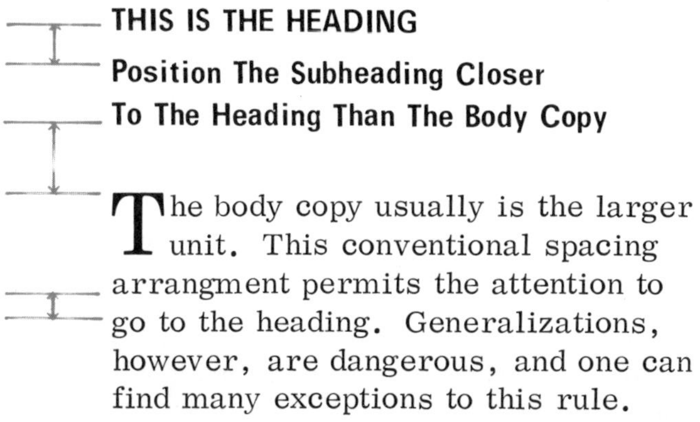

4-13. Type spacing is influenced a great deal by the surrounding graphics. This drawing is a starting point.

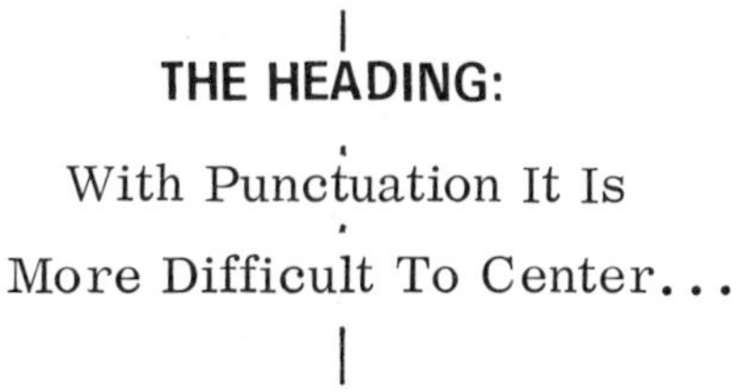

4-15. The visual effect must dominate the literal center—this is the aesthetic of paste-up.

Artwork

Technically everything that comes off the printing press is line art. The reason for this is that printing ink is one solid color, such as black, and it is either on the paper or it isn't. But what about all the photographs you see in newspapers and other publications? They look like tones of gray, not a solid color. While they appear to be tones of gray, they are really made up of tiny black dots—line art—surrounded by white paper. Because the dots are so small and the printed picture is viewed from a distance, they seem to merge and form gray tones. You can easily check this out for yourself with a magnifying glass.

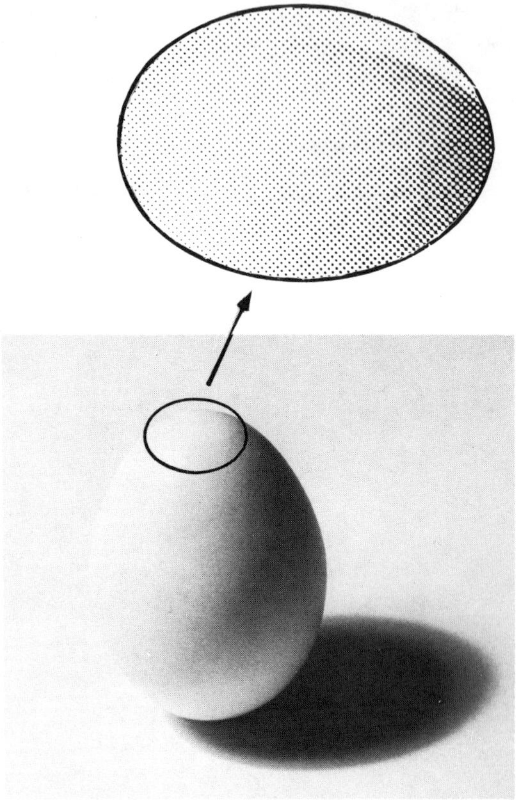

4-16. Dots equal tones for most printed matter. Lifting the top off this egg and enlarging it four times reveal the dot pattern that creates the illusion of tone.

Look at a newspaper photograph in good light with the magnifying glass. Do you see all the dots? The darker the tone, the bigger the dots; the lighter the tone, the smaller the dots and the larger the area of white showing between. This is called a *halftone.*

Halftones can be produced in a variety of forms and styles. All of them employ the same basic dot technique to create the illusion of tone. The original art is continuous-tone like a photograph. The camera produces the dot pattern on the negative by placing a screen between the art and the film at the time of exposure. Halftones must be photographed separately from line art, which does not require a screen. For this reason a paste-up shows the position and size of halftones, but they are usually not placed on the same board with the copy and line art.

A halftone that must be resized to fit the design specification is reduced or enlarged by the camera, and the correct-size halftone negative is combined with the other negatives. Line art can also be resized in this manner but more often it is included on the paste-up flat in the right size.

If line drawings have to be reduced or enlarged, they can be photostated to the desired size, and the prints included on the paste-up. Decisions of this kind are usually made on the basis of cost. The quality of the art should be maintained through several stages of filmwork. If the quality would suffer from this kind of handling, then the line art would be handled as if it were a halftone.

Halftones, on the other hand, are much harder to reshoot, and for this reason the closer the printed piece is to the original, the better the quality will be. As a rule it is much better for the platemaker to shoot from the original halftone art than from a screened print—a print of a halftone that already has the dot pattern. However, a halftone that has been screened can be shot as line, and, if cost is the overriding factor, this is often done.

If halftones have to be reshot, as do screened prints, or Veloxes, it is best to work from a coarse screen. You will notice, if you compare a newspaper photograph to a magazine reproduction, that the magazine halftone is printed on better paper with a finer screen. The fineness or coarseness of the screen is measured by the number of dots to the linear inch. The traditional 65-line screen is more common in magazines. Specific screens are more or less standard for specific applications, although there are variations, depending on the printing process, the paper, and the detail of the subject—an extremely fine reproduction may require up to a 300-line screen.

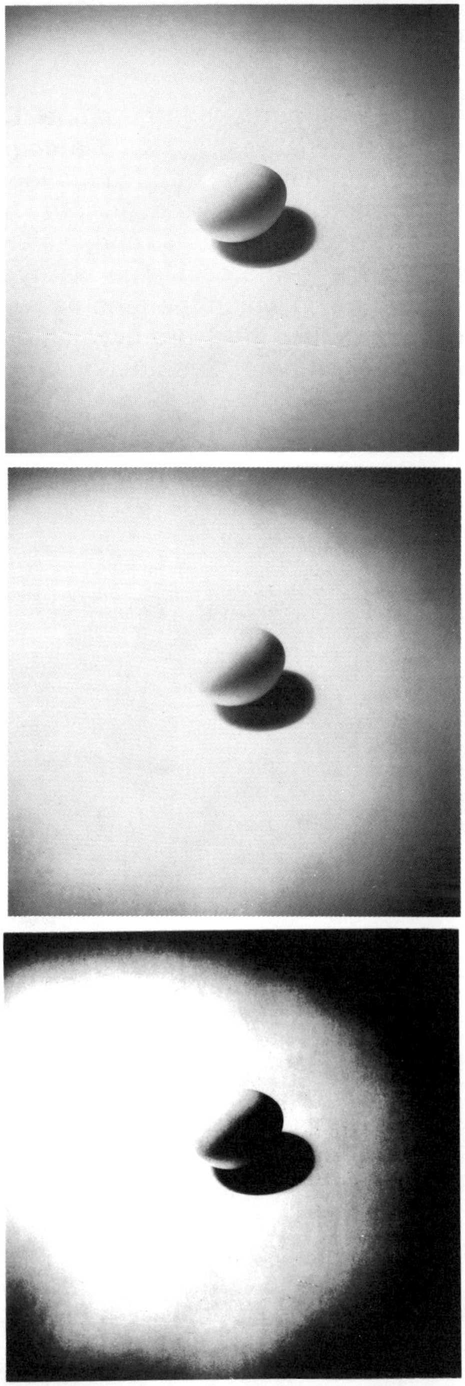

4-17. As a copy is made of a copy, the quality drops. The first copy is from the original; the second copy is a shot of the tiny dots in the first copy—that is, the first copy was used as a screened photoprint; and the third copy was rescreened from the second—that is, copy with a dot pattern was again broken into a dot pattern. This will work only if there is a great reduction in size. Usually a moiré pattern results.

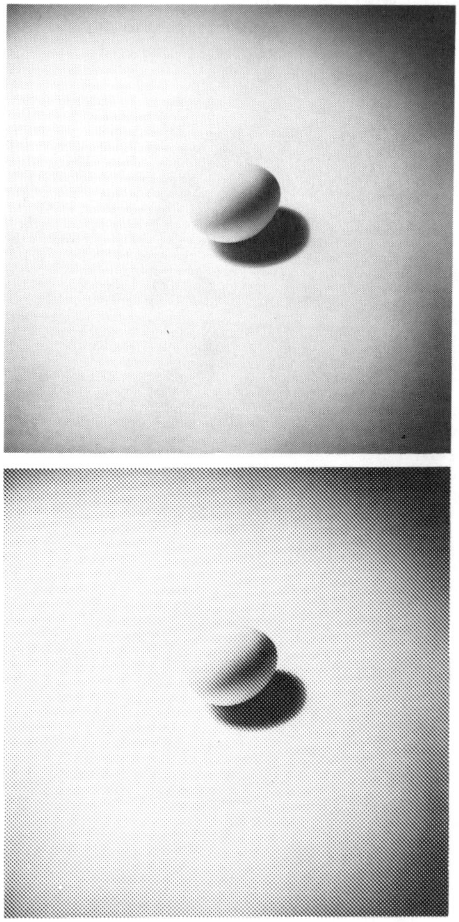

4-18. The eggs show the difference between a 120-line screen and a 65-line screen. The finer screen permits more tonal contrast and smoother gradations and requires better paper.

The most frequently used type of halftone is the *square* halftone. It is the simplest and the least expensive. Another common type is the *outline* halftone, in which the background is eliminated and only the object of interest is printed. For art illustrations, as opposed to photographs, there is the *dropout* halftone, so named because areas of tone within the picture are dropped out. This is typically seen in newspaper reproductions of fashion figure drawings. Dropout halftones can also be combined with portions in pure line. This type, done with two negatives, is called a *combination line-and-dropout* halftone, or combination dropout halftone for short. There are many possible variations for artwork illustrations, and, naturally, the extra work adds to the cost. The last type is the *vignette* halftone, in which the tone background is faded off. Altogether there are five halftone types: square, outline, dropout, combination dropout, and vignette.

4-19. Starting with the original photograph, a halftone can be handled in a number of ways. These illustrations show the four basic art markups and the art that results.

1. The original photograph.

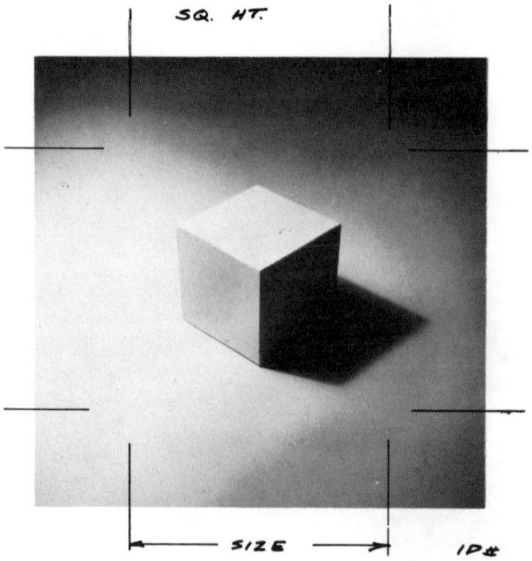

2. Square-halftone markup.

4. Outline-halftone markup.

3. Square halftone.

5. Outline halftone.

4-20. Artwork, as distinguished from photography alone, adds the combination halftone to the list of possible treatments. More than one treatment can be used at the same time. This is a wash-and-line drawing. Notice the strength of the lines in the combination halftone. In the square halftone the white paper holds a background tone.

1. Square halftone.

2. Dropout halftone. It becomes a vignette because of the art treatment.

3. Combination halftone.

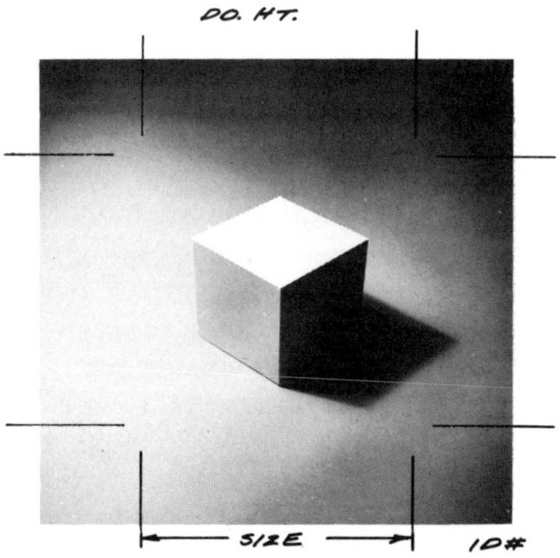

6. Dropout-halftone markup.

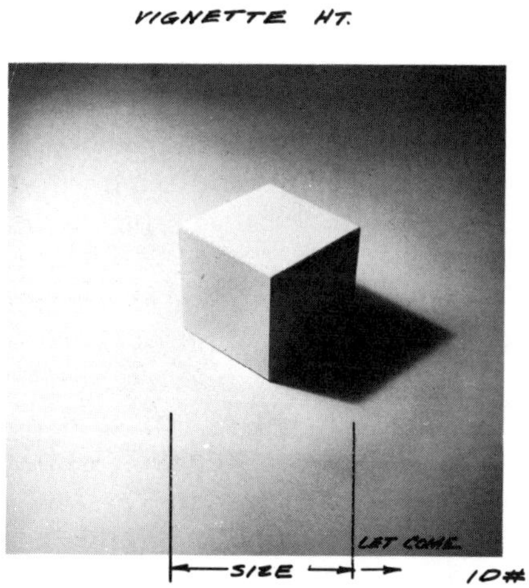

8. Vignette-halftone markup.

7. Dropout halftone.

9. Vignette halftone.

Added to these basic treatments for halftones is the possibility of incorporating reverse negatives of line art, shown in the example in chapter 3 for the heading. These negatives can be used together with a halftone to overprint or reverse out line art. This sounds much more complicated than it is, and in fact, terminology is the basic problem in conveying the proper information to the platemaker.

Fortunately there are certain basic conventions in the presentation of art that help to communicate which treatments are desired. For example, if you want an outline halftone, paint a thick white line around the central object in the photograph presented for shooting. Outline halftones can also be done with a mask, or knockout, as it is sometimes called.

4-21. Art for outline halftones should usually be prepared by the artist. The example at the top is outlined in white paint; the bottom one has an overlay mask, sometimes called a knockout. The advantage of the mask is that the cameraman can make an ideal exposure without dropping out the background dots in the white, and no handwork is necessary.

The paste-up flat also helps make this clear by showing the position and size of the final art. In addition, the art should be marked as an outline halftone, often written "ol. ht." The same instructions are used for the other forms of treatment, with the exception of the square halftone—because it is so common, the labeling is often omitted.

Assembling a Complete Paste-up

Now that you understand how a professional handles body type, headings, and art, you are ready to put them together and do a complete paste-up job. Your paste-up will include all the elements usually found in a publication, a catalog page, or an advertisement.

At the end of this chapter you will find all the material for the paste-up—printed art (use as if it were a photograph), galley proof, photostats of line art, and reverse type—and a rough working layout. The instructions are to make a 5-×-7-inch paste-up; crop the art; make a spelling correction in the last line of the body copy; flap the art; get signed approvals from Tom Artist, Dick Executive, and Harry Client; and have the art picked up by PDQ Platemakers, with written instructions to PDQ to make a negative and ship to Wheat & Chaff Printing via Whirlwind Freight. Start timing the job when you are given the instructions, because travel, accumulating materials, and other details are charged against the job. The instructions and keyed illustrations tell you how to do it.

1. First make sure that you have all the pieces for the paste-up and put adhesive on the proof, the photo, and the photostats. Get flap paper and tissue tracing paper. Use two boards measuring approximately 8½ × 11 inches, which allows room around the 5-×-7-inch format for instructions. If the boards are the same size, they will be easier to wrap and handle and will make a neat package.

2. Starting with the paste-up flat, put the 5-×-7-inch shape in the center of the board and draw the guidelines in nonreproducing blue. Put crop marks in the corners of the art so that they will show on the negative. These marks show the platemaker the page position. Draw a red-outline box for the position and size of the square halftone. (The halftone

isn't really square, but rectangular shapes are also called square halftones.) Label the art—AA—and mark it inside the red key-line box. Remember that red signals the platemaker that the lines are not for reproduction. The red key line will photograph as if it were black, however, and show the position and size of the halftone in the negative.

3. Ink in the rules shown in the layout. The design you are working with allows no variation, so they can be done first if you wish.

4-22.

1. A rough working layout is accurate enough to transfer quickly by making small pushpin pricks at strategic points. Some artists object to this method, because if the pinpricks are not covered by type proofs, the hole will cast a shadow that will be picked up in the platemaker's negative. Be sure to check the measurements against the specifications.

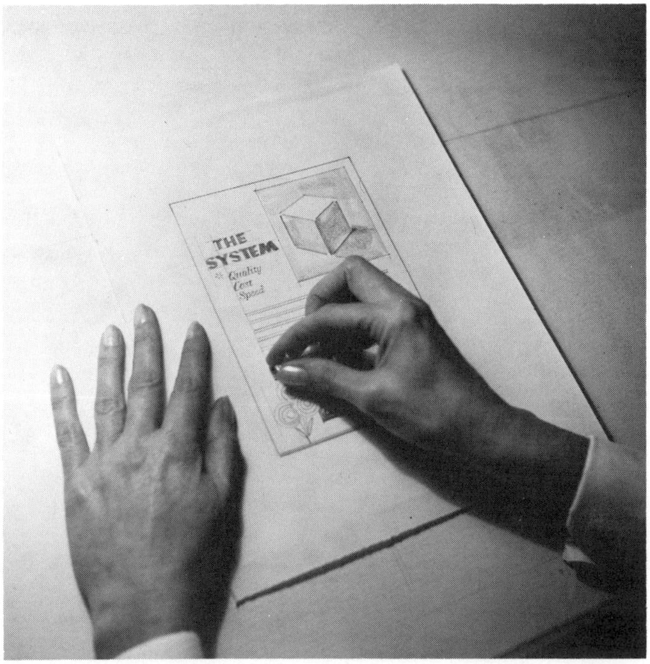

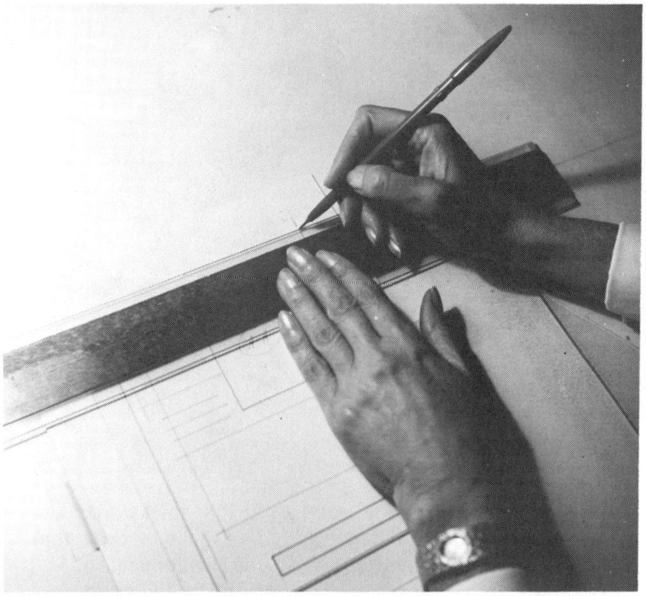

2. This layout calls for a sheet size. It is indicated by placing trim marks outside the sheet area so that they will be trimmed off. Make sure that they are in the proper relationship to the position of the art on the sheet.

3. This design does not allow any variation, so the rules may be inked in first if you wish.

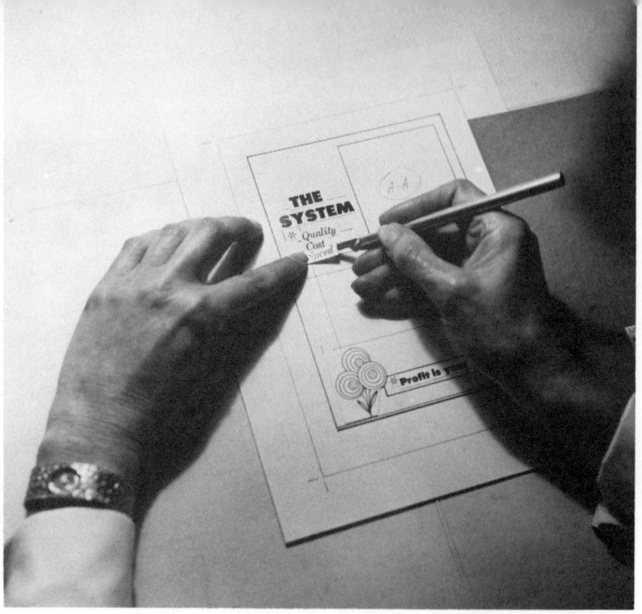

4. The type has been set in such a way that each word must be pasted up individually. The advantage of this method is that the paper areas above and below the type are identical and help in spacing the type. This is preferable to setting the words above one another but with the wrong spacing, which requires more trimming.

5. The body copy runs one line short of the layout—a common problem. Since typeset material does not always conform to the layout, it is a good idea to check before you start the paste-up. The problem is more serious if there is too much copy for the space.

4 and 5. Mount the line art, headings, and body copy. Position the reverse type accurately and trim the edges to form the panel. The last illustration in chapter 3 shows the reverse-negative method for making a reverse panel. It can be used in place of a reverse photostat to achieve the same result, but you have a reverse stat with your material, which saves a negative for the platemaker. If you make a mistake, ink the border of the panel, mount the reverse type in the proper position, and paint the edges with black ink to fill the panel. With everything in position and accounted for, burnish down. Checking everything over should remind you to make the type correction.

6. Tape the tissue or tracing paper to the top of the flat and remove it from the drawing board.

7. Trim the tracing paper.

8 and 9. Mount the flap by taping it to the back of the board so that it can be folded back for shooting. If the flap paper is scored or cut lightly, it will fold cleanly.

6. When the paste-up is complete, mount a tissue overlay for protection and instructions.

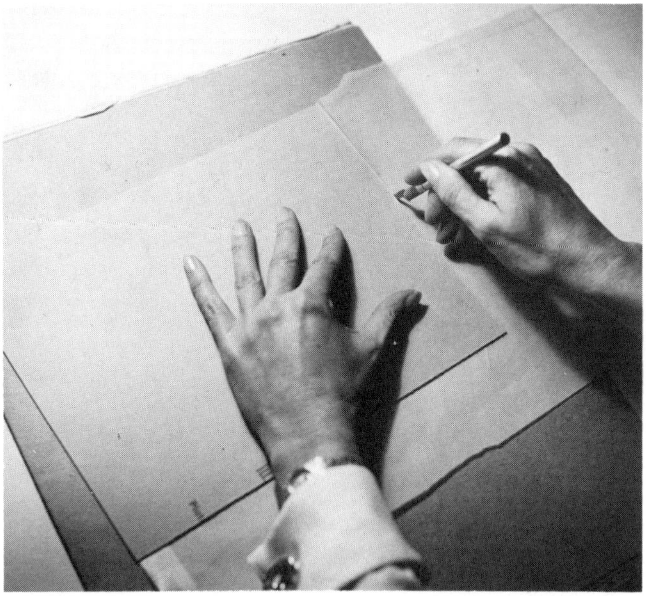

7. This method for trimming the tissue overlay gives an exact fit. Be careful not to cut the mounting board. If the flat is thin, use a straightedge.

8. Cut the flap from heavier stock and leave about a 2-inch extension to hinge over the back. Score the flap the thickness of the flat to position the folds.

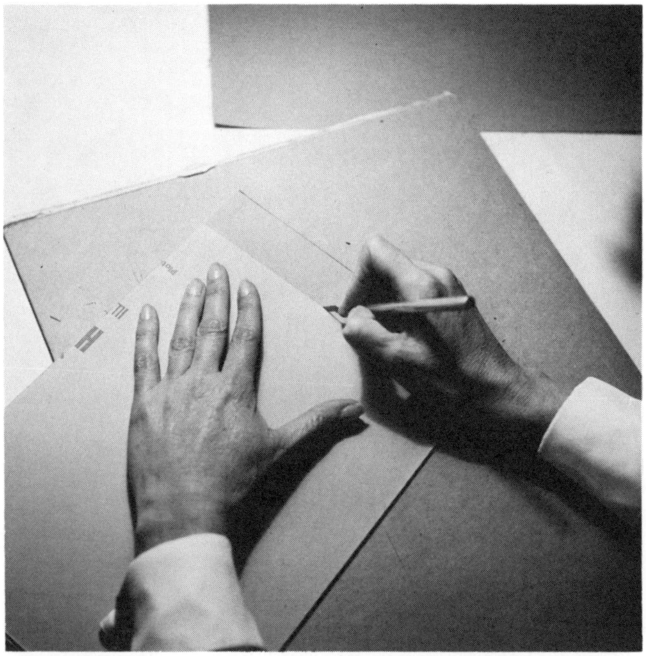

9. With the scores (or light cuts) the flap can be folded over the paste-up flat and taped to the back, allowing it to be folded out of the way for shooting.

10. Check the work to make sure that everything is in place and that it looks clean and crisp. Mark the key-lined halftone.

The following instructions and illustrations show you how to scale the art.

1. Position the other 8½-×-11-inch flat on the drawing board and get the halftone art ready. Trim the art cleanly and mount it in the center of the flat. Lay a piece of tracing paper over the red box for the square halftone on the paste-up flat. Draw the left and bottom lines on the tracing paper, tracing over the red box, and then draw a diagonal line accurately from the lower-left corner through the upper-right corner. Extend it several inches, as in the illustration. Place this tissue over the art and, with the pencil lines for the left and bottom sides, align a transparent triangle to mark the top and right sides.

2. Adjust for position and cropping, keeping the corner of the triangle exactly on the diagonal line to maintain the proper proportion and the angle at 90 degrees to maintain the square. When you have the right cropping, make a pencil dot at the corners just hard enough to leave a slight impression in the art.

3. Here is how to use a Scaleograph.

4. Using the dots and a straightedge, mark the cropping at the edges of the art—in this case, the photo. Make the marks outside the area to be shot. Mark the size across the base, key the photo AA, as on the paste-up, and add any further shooting instructions. Mount the protective tissue and flap as with the paste-up. In some cases it is desirable to explain the cropping on the protective tissue or to make something clearer with a note on the paste-up flat.

5. Is everything complete? Don't forget the type correction. Is everything clean? Fill in the job ticket and get the signatures.

6. When everything is approved, get the shipping instructions, wrap the package, and drop into the pickup box, with the job-ticket label (included with the material at the end of the chapter) filled out. Check your time and you've finished a complete paste-up job.

The organization of steps shown here, while logical, is not a set formula to follow. Most artists work by an association of ideas to the requirements of the design. In some cases certain steps must be completed before others can be taken, yet there is no rule as to their order. Flexibility lets you adapt to various conditions.

4-23.

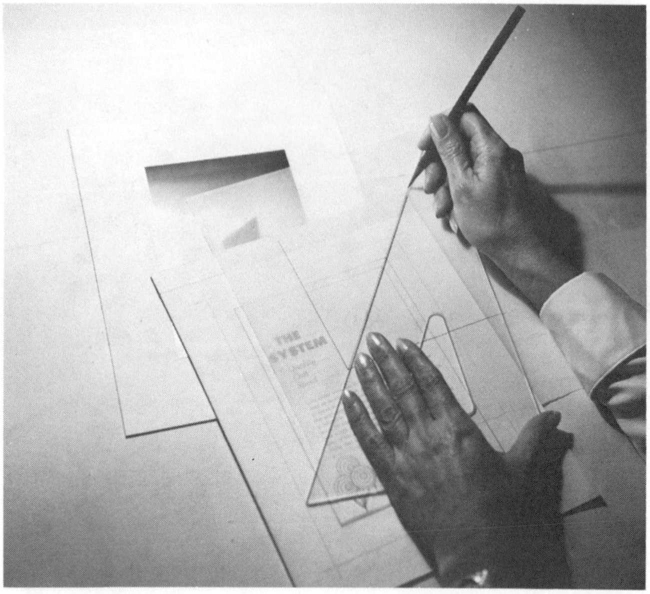

1. Mount the photo and make a diagonal scale on tissue for sizing and cropping.

2. Using the 90-degree angle on the tissue for one corner, move a triangle up and down the diagonal line to make the other corner. By sliding the tissue and moving the triangle up and down, the picture can be framed precisely.

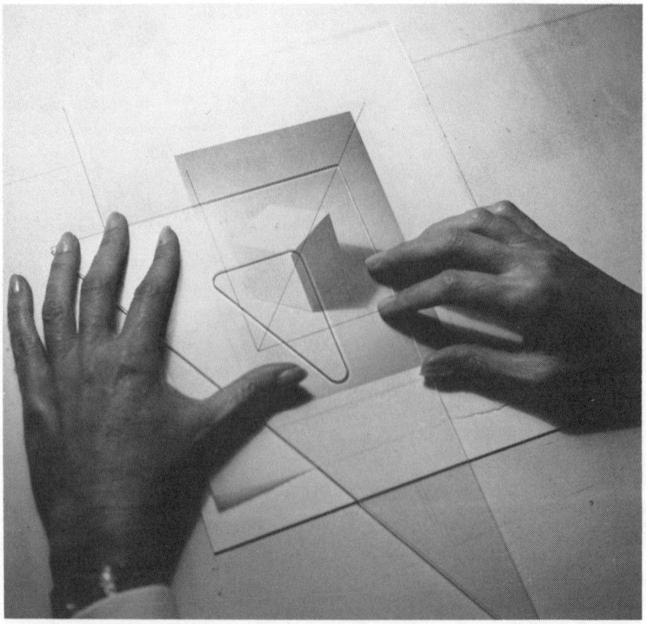

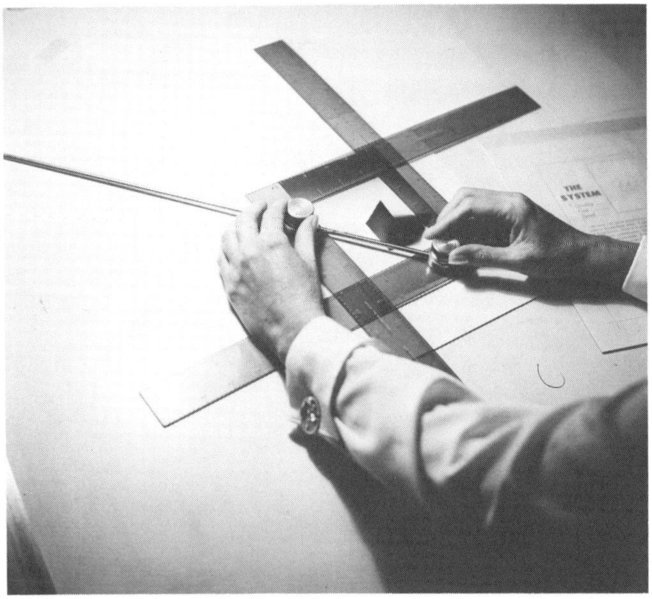

3. This illustration shows how to use a Scaleograph to crop the picture. In principle it works like the previous method. Note that the diagonal (the metal bar) can point in the opposite direction.

4. Using a straightedge to make the crop marks provides a second check that they are in the right place and square to the picture.

5. Check again—did you remember to catch the typo in the last word of the body copy? Use the extra galley to make the correction.

6. Wrap the package for shipping. If you are not sure who will be handling it, add a cardboard stiffener so that the art will not be folded, which would destroy it.

4-24. The clip sheets on the following pages are duplicated at the end of the book.

CLIP AND USE THIS PAGE FOR MATERI[AL] TO DO PASTE-UP

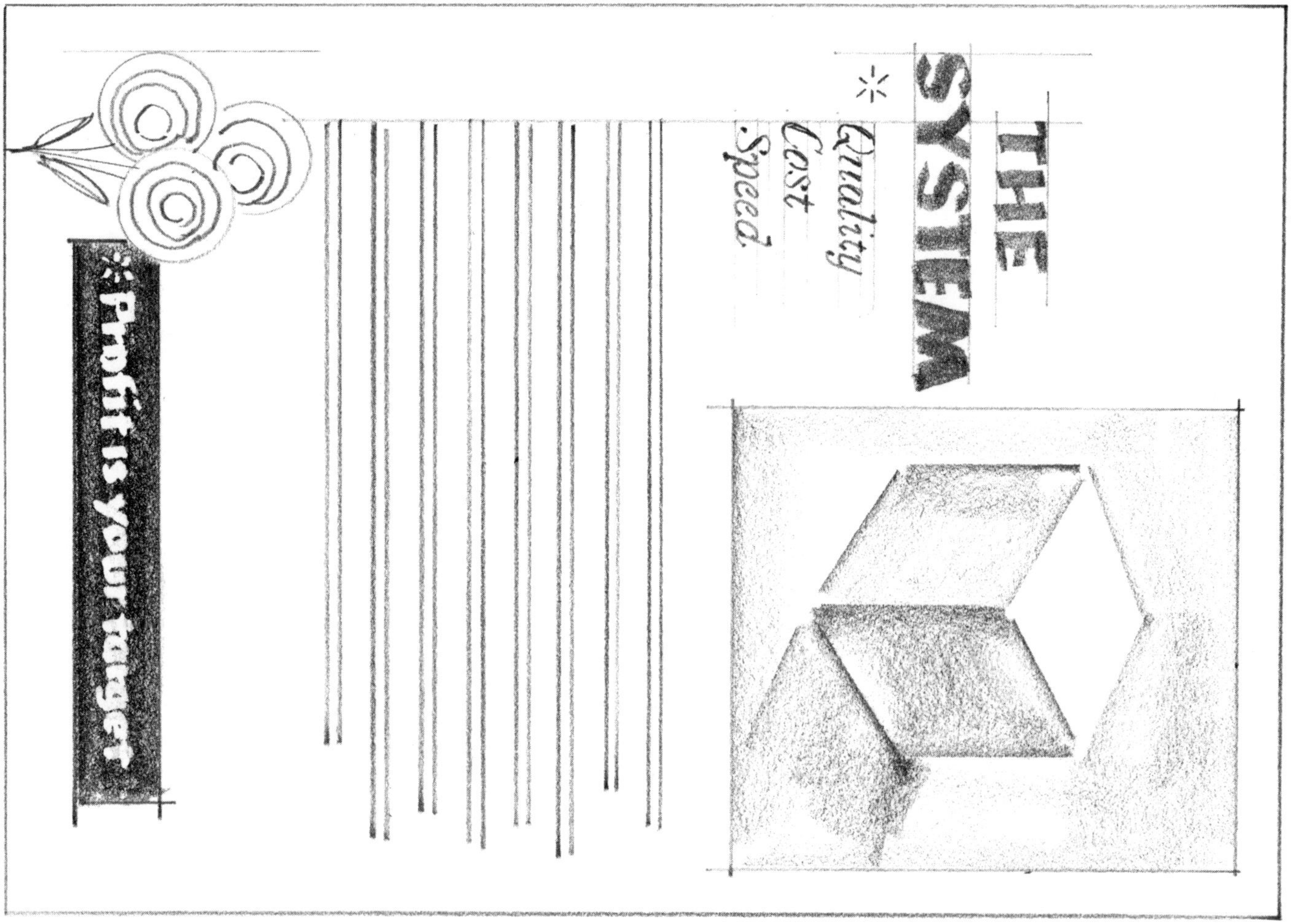

← LAYOUT ON 6 3/16" × 8 1/2" SIZE SHEET

THE SYSTEM

Quality Cost Speed ← TYPE GALLEY

Like many arts put to practical use, Paste-up can be seen as having three sides. It seems to be a balance of three forces. First it is an art form, and quality is the primary consideration. Yet, if it costs too much it becomes impractical. Further, to be practical, it must be available when it is needed. Many say that this might all be summed up by talking about cost, but that is too simple. Sometimes new ideas are the thing that is needid.

← STAT

Like many arts put to practical use, Paste-up can be seen as having three sides. ← EXTRA TYPE GALLEY

← PHOTO

*Profit is your target

← STAT

JOB TICKET →

Quality PASTE-UP COMPANY

JOB TICKET

TO: ______________________

NAME

DATE ______________________

ADDRESS

____

CITY STATE ZIP

SHIPPING INSTRUCTIONS ______________________

____

____

JOB ______________ JOB# ______________

PASTE-UP

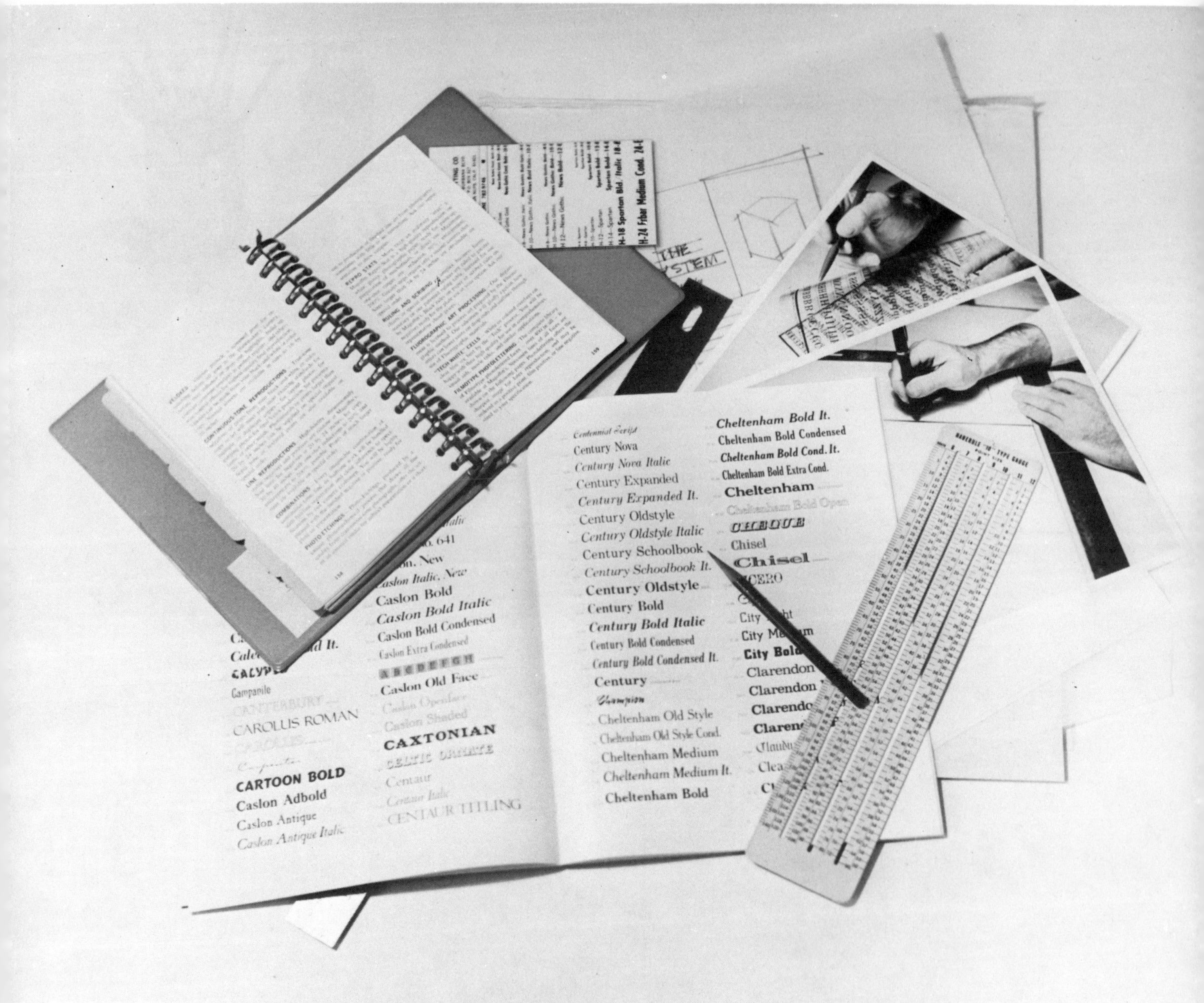

5-1. While the terms "paste-up" and "art assembly" are sometimes used interchangeably, art assembly includes more production functions.

5. Art Assembly

Art assembly differs from paste-up in that it is a more inclusive term: it covers the gathering of art and art materials and maintenance of production flow—contact with type and typesetting and dealing with suppliers such as typographers and photostat houses. The job requires more experience than pure paste-up, and it is found in art departments that have the degree of organization and work flow to require these services in addition to paste-up. It is a position under the supervision of a manager—an art director, a production manager, or a creative director, depending on how the business is organized. The work straddles the art and production departments, with extended responsibilities for parts of each.

The terms "production artist," "mechanical artist," and "mechanicals artist" all refer to art assembly. The difference in meaning between art assembly and production is that production can be entirely deskwork and requires only an understanding of the art process, whereas art assembly is a creative job for an artist. There is a difference between a production artist, who is mainly involved with business, and art production, which requires an artist and is mainly involved with paste-up. A mechanical is basically a paste-up. If there is a difference between a mechanicals artist and art assembly, it is that a mechanicals artist may perform the required mechanical drawing—inking rules, etc.—but no design. The mechanicals artist works to the specifications of an art director, designer, or layout artist who has determined the design within a fairly tight layout. The term "mechanical" may be in contrast to "design."

These terms are basically used interchangeably, but there are slight differences that you should be aware of in looking for a job. Production can mean no drawing-board artwork, and mechanical can mean no design. Assembly or art assembly is involved with art and includes paste-up along with a certain amount of responsibility for the production function. The paste-up flat with the finished art in position is in this context called the assembly or the art assembly. The assembly artist obtains the typesetting and the photostats, produces the required line art and the paste-up, marks up the art for the platemaker, and often schedules the work with suppliers, shippers, and publishers. He or she may also be involved with estimating, costing, bidding, buying, and in smaller organizations billing, bookkeeping, and traffic.

Gathering Art Materials

Paste-up is the stage in the production system at which art materials are gathered. It is the end of the line for design and design preparation, the point at which design becomes finished art. Paste-up is the end point for suppliers to provide art materials that are to be incorporated in the final product and the point at which material developed within the organization goes to outside suppliers.

In every system of art production—unless there is a formal control point for shipping art to printers and publishers—paste-up is the natural focal point at which all the various materials come together. It is for this reason that the paste-up artist becomes a production artist. Many organizations are structured so that paste-up is under production management and not under art management. In another system there is no separate production department: everything concerned with art production is under art management. Depending on the system, the production manager can be business-oriented or art-oriented. The policy regarding the chain of command is determined by the management. The chain of command structures the work flow and places the responsibility. Work flow and responsibility are not static—they change with the needs and demands of the market.

Because of the paste-up artist's position he or she is often given assignments to order materials from particular suppliers and to assemble these materials. Since there is no hard-and-fast line between design and production in this kind of situation, the paste-up artist will gradually be asked to take on more actual design work. This illustrates the method by which the paste-up artist is apprenticed and gradually assumes more responsibility. In another interpretation of the paste-up function and a different kind of production system the lines between departments and the assigning of responsibilities are more definite. Here there are specialists to fill the various functions of art assembly. In other words, an art department can be structured *horizontally*, with a number of artists doing the same thing within a controlling system, or *vertically*, with specialists for each function in the system. In a vertical system art assembly extends to the boundaries of the adjoining specialty—that is, if there were a production specialist who handled traffic and contact with suppliers, these jobs would not be included in assembly. There is an infinite number of variations between the two extremes.

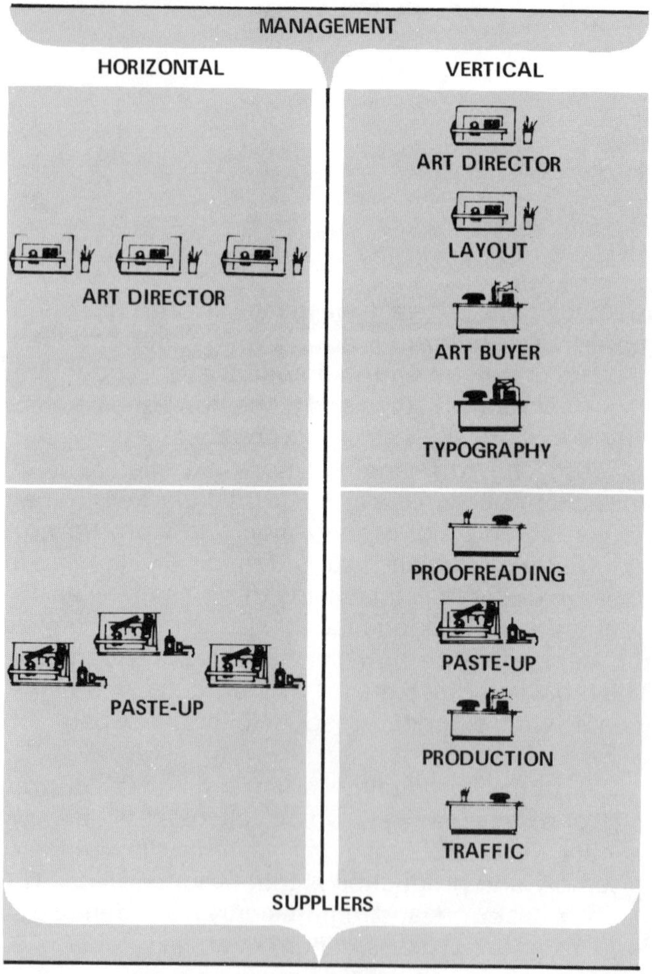

5-2. No two art departments are organized in exactly the same way: some are horizontal—a loosely organized group of artists does a number of overlapping functions—and others are vertical—a more tightly knit group of specialists.

Production systems are structured to an extent by the needs of the product. In the two main categories of advertising and publishing, advertising favors a vertical arrangement, while publishing tends to a horizontal. New tools and materials bring new factors into play and influence the arrangement of these system structures. In a small office a single individual may perform all the functions, including design. Instant printing provides the means to simplify the procedure involved in producing a printed piece, and the increasing number of new materials can be used to shortcut the normal art-production system.

The effects of new materials and methods on art production in publishing are well illustrated by a simple system developed along the lines of business-letter writing, utilizing cold-type typesetting. As with instant printing, the major design considerations are established and controlled by a system of art production. The decisions are made by the paste-up artist at the time of paste-up. The art-production system in effect is used to establish a format, and the format takes care of the concept and the design structure. The paste-up artist arranges the elements according to this structure and adjusts the layout to complete the design.

No art director, designer, or layout artist is involved with the work. The design is established by the format and the format is determined by management. Often the typing and typesetting are done in-house by a typist. The editorial members of the team determine the content and supply all the material. All members work to a clearly defined format but without an actual layout of a specific design. Responsibility resides with the editor.

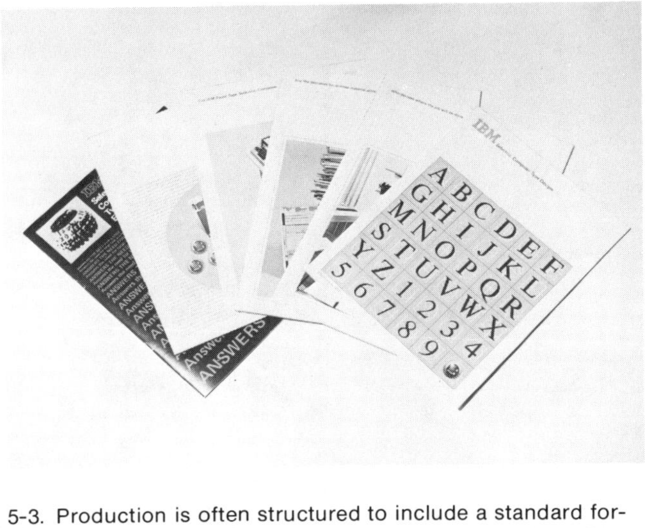

5-3. Production is often structured to include a standard format. This well-designed format for brochures is used by IBM to advertise their type composer.

The economies possible with this kind of a system make it feasible to publish material which would be economically impossible with conventional systems. Publishing houses are adopting variations of this system by the thousands. Some of the variations include an art department with design and paste-up. According to the budget, the art-production system can use art assembly with a format or design and paste-up. Paste-up is the basic function that cannot be dispensed with—there has to be some method of assembling the graphic material for printing!

The question for the paste-up artist is: how much design and how much format? Every piece in a paste-up involves some decisions. The looser the format, the more the decisions become involved with design. The same thing applies to layout: the looser the layout, the more decisions remain to be made. At what point does this become design? There is no hard-and-fast answer to this.

The importance of an aesthetic sense and design quality has already been pointed out. In the publishing production system just described it is easy to imagine the design demands that can arise at the time of paste-up. Situations often develop that call for more than "fitness," especially in the absence of a tight layout. There can be design problems, for instance, if the elements do not fit the allotted space and the format does not allow for this eventuality. Enter design and layout.

The same process can occur in other ways and from other causes. Late ads that need layout may end up being done at the paste-up board to expedite production. A rush of work can spin off some layout to paste-up or result in very loose layout roughs. Last-minute changes can have the same effect.

In a situation that involves considerable expense, such as some advertising work, sloppy work would never be tolerated and this problem could never occur. In production systems that produce large quantities of material this can happen due to sheer volume. The fact that paste-up is the final step in the production process implies that the work is completed here—it must be camera-ready. This includes everything, whether it is incomplete, late, a mistake, or a change—all the unexpected problems that have to be solved to complete a successful design.

How these situations are resolved depends on the policies of the management and the art department. The paste-up artist is not supposed to have responsibility for design but may inevitably inherit some design. Art assembly does include some responsibility for design.

Layout

The layout is the blueprint for the finished piece—the artist follows it in doing the paste-up. Layouts vary widely in detail of execution: they can be comprehensively rendered designs that look like finished pieces or simply designated formats with understood boundaries and forms but no actual rendering.

A comprehensively rendered design is called a *comp* or a layout comp. The work is done in ink and paint, and sometimes finished art such as photoprints is included to make the design as representative as possible. These layouts are painstakingly rendered, often works of art in themselves, and costly to produce as a result. The usual reason for making a comp is to explain ideas to clients clearly enough to eliminate any misunderstanding. The more experience and professionalism

5-4. All layouts share a common function, but there is a wide variety of presentation. On the left is a rough layout; in the center is a working layout; and the layout on the right is a rendered version, which is sometimes matted for presentation. A comprehensive layout is difficult to distinguish from the final job.

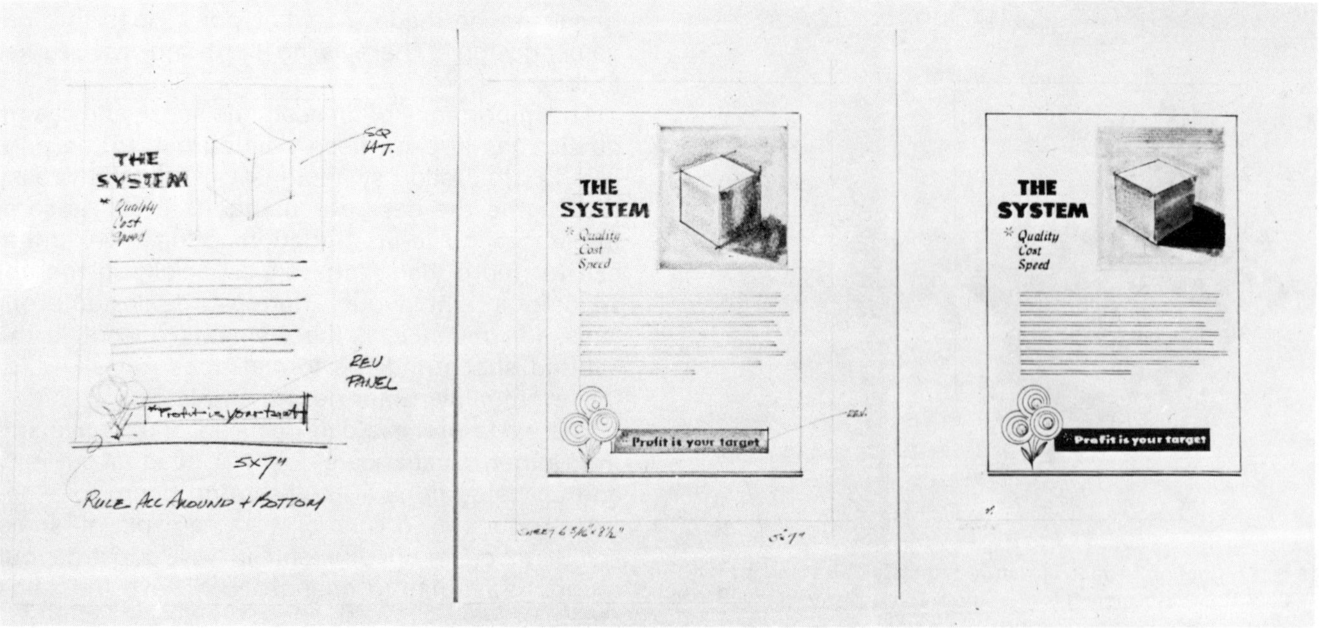

the people involved have, the less need there is for a comp. When the paste-up artist works from a comp, the paste-up should be as visually attractive as the comp, because it can also be the subject of client scrutiny as well as serving as camera-ready art.

A *layout rendering* shows all the elements in proper size and position, but they do not look exactly like they will in actual reproduction. The work is done in ink, pastels, and graphite. The elements are indicated in a stylized way to give the impression of the final printed result. Layout renderings are often very attractive, with all the qualities of a sketch. They are also used for presentation, especially in publishing, since the work is mainly done by professionals.

The usual working layout is a *rendering on tissue*. All the elements are positioned and sized accurately, but the rendering is usually a sketchy impression. All the actual design problems are solved, but little time is spent to make the layout look like the final result. If problems do occur in pasting up from this kind of layout, you should ask what was intended, because an omission is most likely an oversight.

Less explicit layouts are called *roughs*. They are not necessarily drawn mechanically and can vary from tremendously accurate freehand sketches to a few squiggles or a small thumbnail-size sketch with dimensions written alongside that define the format of the design. Roughs are usually accompanied with verbal instructions about what is needed.

The defined *format*, with no sketch, can work as a layout. It is an effective work-saver if it is well conceived and designed. It works especially well if a designer goes over the elements and handles the problems before they come to paste-up.

The paste-up artist often works with a layout artist, designer, or art director. These terms can be used interchangeably and often are, with "art director" the preferred title. In a vertical production system they indicate an ascending order of responsibility: the layout artist or designer is responsible for design work, while the art director is responsible for the overall graphic concept. Freelance designers often do both jobs.

Some management policies do not actually delegate conceptualization to an art director. The art director's work is limited to design and management of the art within the framework of a given and predetermined concept. The paste-up artist's work in this system can come from management as well as the art department, and designers often do their own paste-up as part of the design process.

This system is being adopted more and more, with paste-up directly assigned by management along the lines of an implicit format. The paste-up artist is the only artist, and while the conceptual structure is supplied, the actual layout is not. Graphic and layout considerations within the basic format must be worked out by the paste-up artist, who is doing art assembly and hopefully layout design.

Copy

It is the designer's responsibilty to specify the type and typesetting for the copy in the design. The designer or layout artist often marks the specifications on the working layout, and in the assembly the paste-up artist marks up the copy from the layout and sends it to the typographer for typesetting. The paste-up artist may also handle the galley corrections and approvals for the camera-ready type used in the final paste-up. Who does what depends on the production system.

Since paste-up is inevitably involved with the finished copy, the paste-up artist should know how typesetting is handled. In larger art departments the actual selection of type and the contracts with typographers are handled by the production manager, and the paste-up artist only needs to know which pickup box the material is put in. In smaller art departments the buying and contracting are usually handled by management—often the specification and contact as well—leaving the paste-up artist to spend more time at the drawing board. Even here, however, the paste-up artist passes the galley corrections along to the typographer.

Methods of typography have broadened considerably recently, and typesetting is often computer-manipulated nowadays. It is cold-type typesetting, either photocomposition or strike-on typography.

Some or all of it may be done in-house as part of the art department, and some companies need artists who can operate a type composer. This is beyond the range of paste-up—it involves typing skills, knowledge of the composer, and type-composition ability.

Out-of-house typesetting is either cold-type composition or hot-metal typography, the two basic methods. It is unlikely that the hot-metal linecaster would be installed on the premises, because it is heavy machinery and requires molten lead to operate.

The galleys resulting from each typography process are different. The paste-up artist should recognize strike-on cold type in particular, because it requires careful handling and is usually sprayed with a fixative to minimize the possibility of smearing. You should also be familiar with transfer type, since it may be used for a portion of the typesetting.

5-5. Different methods of typography utilize very different kinds of machines. Above is the Linotype linecaster (hot type), with a rack of font trays in the background. To change typefaces, the tray on the machine must be changed. At the top is the IBM (stand-alone) Selectric Composer (cold type). Changes of typeface are made by mounting a different-font type ball, shown to the left of the composer. The Harris Fototronic 600 at the right is a photocomposer (cold type). Fonts are changed automatically by keying the code for a new font negative to move into position.

However the typesetting is handled, the copy must be marked up for the typographer. To produce the typesetting, this information must be provided: (1) the size of the typeface in points, (2) the type body size or leading in points, (3) the typeface design by name, (4) the measure or width of the line in picas, and (5) the configuration of the layout—flush left, ragged right; centered; flush right, ragged left; justified; or random.

Instructions for the text you are reading are written on the copy as: 10/12 Helvetica × 20 picas, justified. The appropriate copy is bracketed, with a line running from the bracket to the specification. Every piece of copy has to be specified. In the example "10" refers to the typeface size (1); "12" refers to the body size it is cast on for a leading of 2 points (2); "Helvetica" refers to the name of the typeface (3); "20" is the line measure in picas (4); and "justified" is the configuration (5). Leading, measure, and configuration can be omitted if they are not needed—for display type, for example.

If the layout is sent to the typographer along with the copy, only the size and name of the type need to be specified, because the rest of the information is on the layout. The typographer does need to know whether to float the copy or to set it in the position shown on the layout. Normally all typesetting is floated, and the paste-up artist positions the material. Typographers get more than $20 an hour for makeup, or positioning type in metal. Almost the only time it becomes necessary for the typographer to do makeup is for tabular material and charts, since the pieces are too small to handle easily on paper. Setting this kind of material in metal makes it possible to adjust the spacing quickly, but the majority of typesetting is a float.

Copyfitting is usually part of layout design. It is not a requirement for paste-up, but you may have to fit a small block of copy into a particular space, especially if you are working to a format.

Copyfitting basically involves counting the characters, punctuation, and spacing in the manuscript and then calculating the space that this material would take up in a particular size and style of type. The problem is to fit the manuscript material into the size and configuration of the layout design.

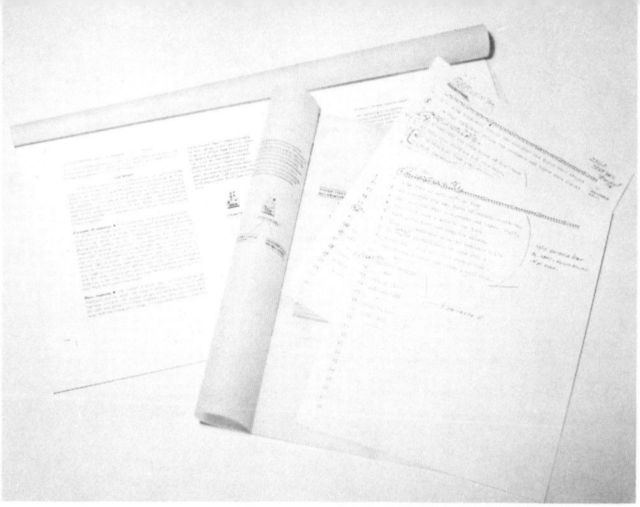

5-6. All typesetting systems work to a similar basic markup, but additional coding is required with an automated system. It is added in this example above the copy. Underneath the copy is the layout, and under that the resulting type, assembled with line art on a paste-up flat.

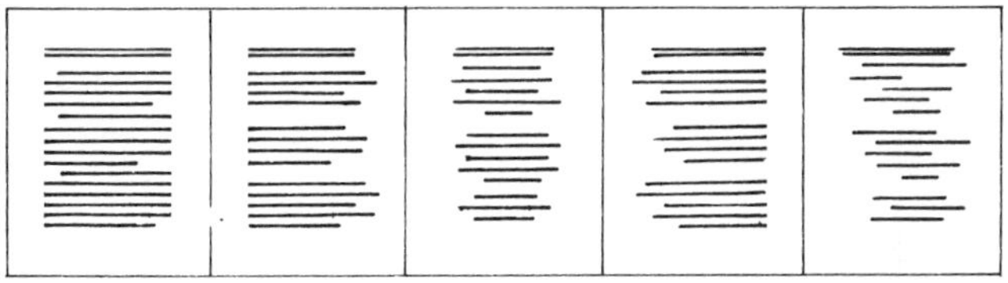

5-7. With all the diversity of typesetting techniques and complex, sensitive machinery, there are only five basic patterns.

TYPE SPACING CAPABILITIES		(X IS NORMAL SPACING)
Letter Spacing	Word Spacing	LINE SPACING
Letter Spacing	Word Spacing	LINE SPACING
Letter Spacing X	Word Spacing X	LINE SPACING X
Letter Spacing	Word Spacing	LINE SPACING
Letter Spacing	Word Spacing	LINE SPACING

5-8. Letterforms are exact and consistent—there are no variations in a given face. Copyfitting is done through minute adjustments to the spacing around the letterform. With cold type the negative as well as the positive side of ideal spacing can be used. Metal (hot) type requires very expensive hand tooling to do this, and space is usually added only to fit copy. Good typesetting normally uses word spacing only to make the copy fit and keeps the letterforms in the most readable relationships.

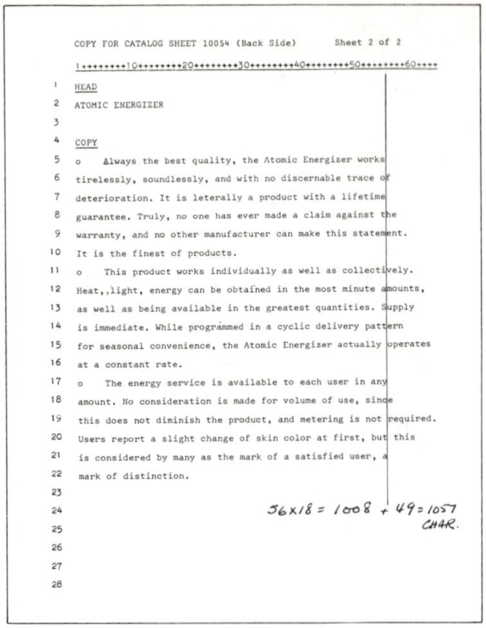

COPY FOR CATALOG SHEET 10054 (Back Side)   Sheet 2 of 2

1++++++++10++++++++20++++++++30++++++++40++++++++50++++++++60++++

HEAD

ATOMIC ENERGIZER

COPY

o Always the best quality, the Atomic Energizer works tirelessly, soundlessly, and with no discernable trace of deterioration. It is leterally a product with a lifetime guarantee. Truly, no one has ever made a claim against the warranty, and no other manufacturer can make this statement. It is the finest of products.

o This product works individually as well as collectively. Heat,,light, energy can be obtained in the most minute amounts, as well as being available in the greatest quantities. Supply is immediate. While programmed in a cyclic delivery pattern for seasonal convenience, the Atomic Energizer actually operates at a constant rate.

o The energy service is available to each user in any amount. No consideration is made for volume of use, since this does not diminish the product, and metering is not required. Users report a slight change of skin color at first, but this is considered by many as the mark of a satisfied user, a mark of distinction.

56x18 = 1008 + 49 = 1057 CHAR.

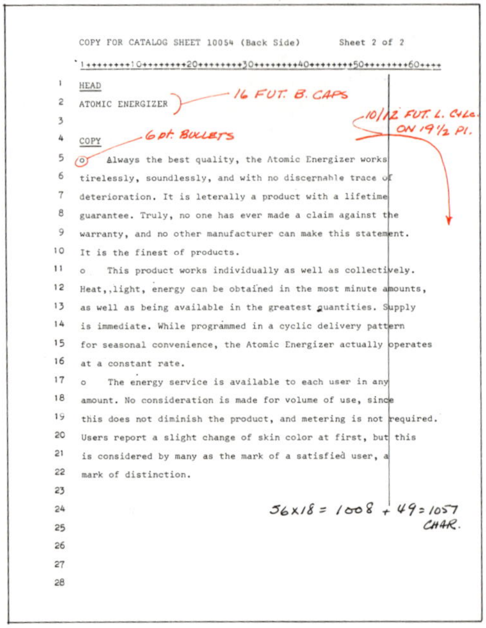

COPY FOR CATALOG SHEET 10054 (Back Side)   Sheet 2 of 2

1++++++++10++++++++20++++++++30++++++++40++++++++50++++++++60++++

HEAD

ATOMIC ENERGIZER

COPY

o Always the best quality, the Atomic Energizer works tirelessly, soundlessly, and with no discernable trace of deterioration. It is leterally a product with a lifetime guarantee. Truly, no one has ever made a claim against the warranty, and no other manufacturer can make this statement. It is the finest of products.

o This product works individually as well as collectively. Heat,,light, energy can be obtained in the most minute amounts, as well as being available in the greatest quantities. Supply is immediate. While programmed in a cyclic delivery pattern for seasonal convenience, the Atomic Energizer actually operates at a constant rate.

o The energy service is available to each user in any amount. No consideration is made for volume of use, since this does not diminish the product, and metering is not required. Users report a slight change of skin color at first, but this is considered by many as the mark of a satisfied user, a mark of distinction.

56x18 = 1008 + 49 = 1057 CHAR.

5-9. The essential information for marking up copy is shown here. With the use of cold type universal names and standards for typefaces have been obscured. Many typefaces have been copied with slight changes and given new names—one typographer's Venus type may be similar to another's Futura face, etc.—so identify the type design with the name that a particular typographer uses.

To calculate the space for copy material, draw a light line after the shortest line in the manuscript. Count the number of characters, punctuation marks, and spaces in this line and multiply by the number of lines. Then count the number of characters beyond the line. Look up the number of characters per pica of the desired typeface size in a type book and multiply by the number of picas in the measure. Divide the number of typeface characters per line into the number of characters in the copy you counted. The result will be the number of lines of type. Measure this against the layout to check the fit. To increase the depth, the lines can be spaced apart with more leading or the typeface can be figured one size larger. To decrease the depth, figure the next smaller size of type or reduce the leading.

These different alternatives make the point that there is no simple way to size type—it is an art. A type book that supplies the number of characters per pica probably includes a method of copyfitting in the introduction. It is proportional, and the amount of type is usually smaller than the manuscript. What makes exact casting impossible is the varying proportions of the typeface letters as compared to the typed manuscript, with one unit of space for every character and word space. Fitting a small block of copy with the method supplied above will give an accurate line count.

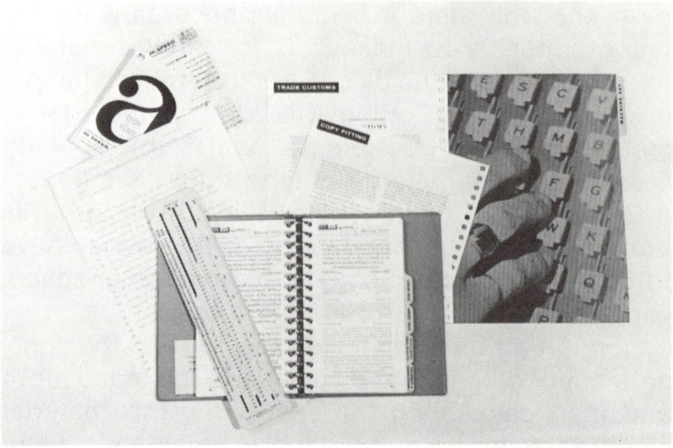

5-10. Additional information can be obtained from the typesetter's type-specimen book. Special rules and typeface finders are also available.

Art

Handling the art is one of the main functions of art assembly. It involves bringing together into one package all the art needed for a particular job and often requires efficient handling of many pieces of art for different jobs at the same time. Art asembly includes proper identification of the art, determining how the art will reproduce, and writing the specifications. The job requires an understanding of the camerawork involved and an ability to make economies by simplifying the production techniques or work flow.

Art markup is different from type markup, but it accomplishes for the platemaker, cameraman, and photo stripper the same purpose that type markup accomplishes for the typographer. Art markup identifies the position of the art in the job and the job to which it belongs. It specifies the size of each piece and the technique to use. If each job is handled individually, as opposed to production systems that budget many jobs on one billing, the art specifications may also include a job ticket that spells out what the job is, to whom it should be sent, and to whom it should be billed. It can include other pertinent data such as purchase orders, order numbers, and related bookkeeping information.

While every job has different individual requirements for markup, there are basic specifications common to all, such as stripping negatives. The unique requirements can be shown on the tissue overlay with a drawing of what you want. The specifications can be noted on the paste-up flat. The basic instructions for handling art are: reverse, overprint or surprint, paint up, and tone panel. *Reverse* means to use the negative image of the art; *overprint* or *surprint* means to combine the image on the overlay with the image underneath it so the two print together; *paint up* means to complete a line or panel where it butts against another piece of art; *tone panel* means to screen the line-art copy to a gray tone specified as a percentage halftone—or percentage of black—which equals 100%.

The usual convention is to do the paste-up flat the same size as the printed piece. If this is not done, the correct size needs to be given. Art prepared separately from the pasteup flat for stripping to position needs to be specified. If it is the same size as the reproduction, it should be marked "same size" or "s.s." If neither the paste-up flat nor the art are the same size, the art must be sized to the final size. It can be difficult to give a propor-

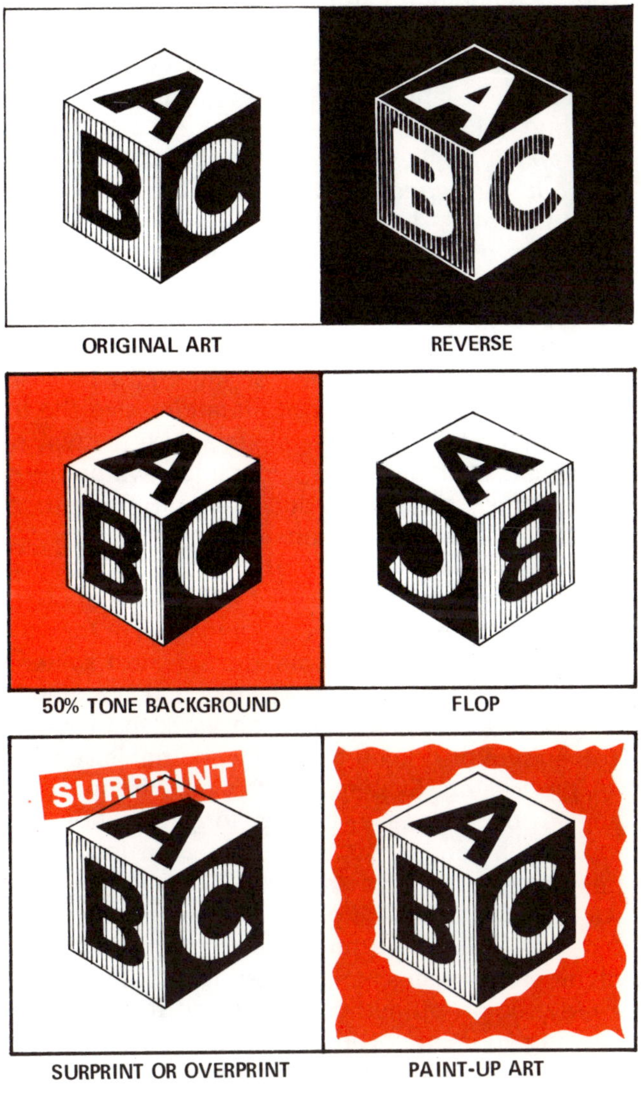

5-11. This is a summary of line-art markups for lithographic film. To make a surprint (bottom left), the art is done on an overlay in proper position for a second exposure (double burn) on the printing plate. To make a paint-up edge (bottom right), a color or tone panel is shown on the art for the platemaker to finish and butt. It enables all the art to be placed on one flat so that the stripper can fit his own negatives for an exact register.

tion of a proportion, which is why the paste-up flat is usually done in the same size.

Art for stripping needs identification and technique specification as well as size. It is identified by lettering or numbering the art to correspond to the lettered or numbered position on the flat. The art technique can be line or halftone. Line art can be *reversed*, *screened*, or *flopped*—the bottom surface of the negative is placed at the top. Halftone art, explained in chapter 4, can be square halftone (usually not marked because the cropping explains it), *outline*, *dropout*, *combination-dropout*, or *vignette.* Markup abbreviations are "sq. ht.," "ol.," "do.," and "comb."

Check your art instructions and try to anticipate ambiguities. What you have so clearly in mind may be expressed in such a way that other interpretations are possible. For instance, "reverse" can apply to the type, the panel, or the entire piece of art. To prevent this kind of situation, give the instruction on the flat and again in a different way on the tissue. This gives the cameraman two points of view to help make out your intention. Be sure, of course, that the instructions coincide!

Wherever you work, there will be a system of some sort to handle the flow of work—files or boxes or bins for incoming and outgoing work or for work at various stages of completion. The best systems show what stage the work has reached at a glance, providing a physical check on work in progress. You should also keep track of work that is sent out for processing. Systems for this vary from carbon copies of purchase orders to production-flow charts that check off each step of a job. Some people seem to have an excellent memory and always know where things are. Remembering helps, but for a volume of hundreds of pieces in dozens of different places it is better to organize some nearly foolproof method of keeping track—and of keeping yourself covered! You can do this by creating checkpoints at certain stages of the work for making an accounting. This need not be complicated—in fact, the simpler, the better. Pick stages that harmonize with the flow of the work, such as a deadline for ordering all the material or for having it all back in house. Better-organized work is easier.

You must constantly be aware of the reproduction quality of the art you are handling—always check the reproduction of tones, lines, and inking. The first proofs are always pored over by everyone concerned. Printers know this and send the best copies. Since there is so much art and craft involved in the process, you have to learn by observation how thin lines can be, how dark or how light, and how a given process affects the art. Then, when the art reaches a critical stage, you can get a better result by making the job easier for the people who produce the work.

A pressman may labor over a critical hairline register when it is hardly necessary for the success of the design. A cameraman can struggle to hold tones that, as far as the design is concerned, might as well be dropped out. A stripper can have a terrible time getting a fit with an overlay when the art can just as easily do without the extra negative.

It is impossible to make rules for this kind of judgment because everyone works differently. Humidity, temperature, conditions, and temperament all play a part. Colorwork adds subtleties of lighting and background color to further increase the variables.

5-12. The importance of fitting filmwork for exposure to the printing plate can be seen here. It is more than a matter of position—wide variations can be achieved with a different exposure or development of the film. The original (left) was made thin (center) and heavy (right) by film manipulation alone. Inking also produces different images.

Color

The change to offset printing in the graphic industries has increased the use of color by making it less expensive. The paste-up artist seldom deals with full color or the four-color process, but he or she will very likely prepare two-color art. Process color is more complicated than two-color printing because it requires four printings: one for yellow, one for red, one for blue, and one for black. When these colors are printed on top of each other in the halftone process, almost the entire color spectrum is presented. The effect is of full color, but certain shades of green, blue, gold, and brown are impossible to reproduce with this technique and require additional color printings.

Scanners are used to produce color separations from color-photo transparencies. They separate the tones into primary colors suitable for the four-color printing process. The negatives are produced in minutes, have excellent quality, and cost less than ever before. Two-color work is even less expensive and is used much more often.

Two-color work is treated as if it were two separate printings of black. The actual colors may be red and black, but the art looks like black and black. The lighter color is printed first, and the darker color, usually black, second. The art is prepared with the black on the paste-up flat and the color on an overlay. If the art is black type on a red panel, for example, the type would be in position on the flat, and the red panel in position on the overlay. The same processes and techniques used to reproduce black apply to each level of a two-color paste-up, but the levels are independent of each other since each becomes a separate printing plate.

The tissue overlay on a two-color paste-up should indicate the desired effect as well as the instructions. If the elements in two-color work do not touch or overlay each other, all the work can be on the paste-up flat, with the tissue explaining which elements go on the color plate and which on the black plate.

Production Specialties

It makes good sense to be on the lookout for ways to save money. One way is to use smaller and fewer negatives by grouping the art. Avoiding an overlay saves materials. Eliminating problems of registry in the mechanics of a design saves labor.

Simplicity is the keynote. The simpler the paste-up, the easier it is to do. The result is economies in production. It can require some hard thinking to find the easiest route to a desired end without compromising the quality of the design. There is no status quo, because materials, techniques, and design ideas are always changing.

The paste-up artist is concerned primarily with preparing and executing the art for the design. Broader considerations of technique, production system, and suppliers are usually handled by management or the production manager.

The paste-up artist can specialize in production. Expertise in this area leads to the position of production manager. As decisions become broader in scope and the size of the production department increases, less time is spent at the drawing board and more at a desk. Large art departments have divisions of production management, such as production supervisors and production foremen.

The scope of the work, while it still is primarily concerned with art, also becomes involved with the business aspects of handling the art. It includes bidding, cost accounting, management of work flow, and determination of procedures. Management of personnel is part of the picture, and the production manager coordinates the flow of artwork in the same way as any other kind of product.

The product is art, but the problems and solutions are common to any business situation and business training is very useful. Because of the specialized nature of artwork, a production manager is often drawn from the ranks of the production artists. It is harder for a business specialist to learn about art than for an artist to learn the practices of business!

To buy and sell art, to produce art commercially, is a business activity. The mystique of the artist tends to obscure this fact, coloring it with other values gained from working in art. Paste-up artist, specialist, art director, production manager—all must function as businesspeople as well as artists. Working within the structure of an organization hides the artist from this awareness to some extent, but a freelance artist confronts it daily. The production manager is involved with art only from the business standpoint—as an artist, the production manager is all business.

How to do Two-color Assembly

For this example assume that you are working with a going art department and production system. This environment provides the material for you to assemble into a two-color catalog sheet. This material is supplied at the end of the chapter. For the sake of the example say that catalog sheets are put out periodically—because with a new category of project there would very likely be more assistance and guidance from management. This catalog sheet is one of many that are put out every three months and tied to corporate advertising and promotional policy.

5-13. The three pieces of art at the top left (a flat and two overlays) are what the platemaker has to work with. Depending on how these are marked up and handled, a wide variety of results can be obtained. Color increases the number of possibilities.

5-14. The two-color catalog page. The art specifications and markup will produce this result on a single, separate 8¼-×-9¼-inch sheet. It is shown here at reduced size. The front of the page shows the black-and-white reverse in its proper printed relationship.

NEW
ATOMIC ENERGIZER
PURIFIES WASTE
PRODUCES FOOD
ACTIVATES PEACE
FOR THE BEST IN PURITY, IN HEAT, AND LIGHT
The Atomic energizer works equally and without regard to the quantity used. A small portion of light has the same quality as a large portion. This is also true in regard to heat, which may be regulated by changing positions of solstice and latitude. Dormant mentality is awakened, and purity remains at a constant state of perfection.
Activistatom • DIVISION OF TECHNORESEARCH CORPORATION
ACTIVISTATOM COMPANY, • H20 PACIFICA PLACE, • TECHNOLOGY, EARTH AD2000, • (LUV) SURVIVE

ATOMIC ENERGIZER

• Always the best quality, the Atomic Energizer works tirelessly, soundlessly, and with no discernable trace of deterioration. It is literally a product with a lifetime guarantee. Truly, no one has ever made a claim against the warranty, and no other manufacturer can make this statement. It is the finest of products.

• This product works individually as well as collectively. Heat, light, energy can be obtained in the most minute amounts, as well as being available in the greatest quantities. Supply is immediate. While programmed in a cyclic delivery pattern for seasonal convenience, the Atomic Energizer actually operates at a constant rate.

• The energy service is available to each user in any amount. No consideration is made for volume of use, since this does not diminish the product, and metering is not required. Users report a slight change of skin color at first, but this is considered by many as the mark of a satisfied user, a mark of distinction.

Equinox . . . . **No Charge**
Summer Solstice . . . . **FREE**
Winter Solstice . . . . **FREE**

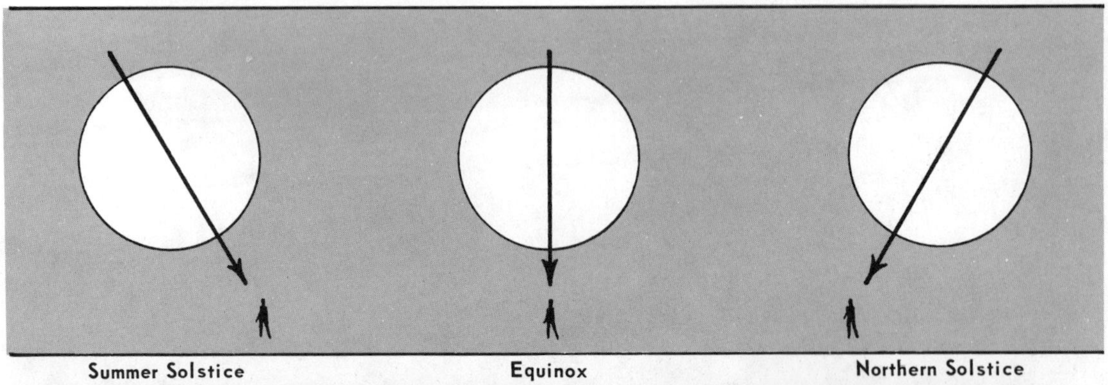

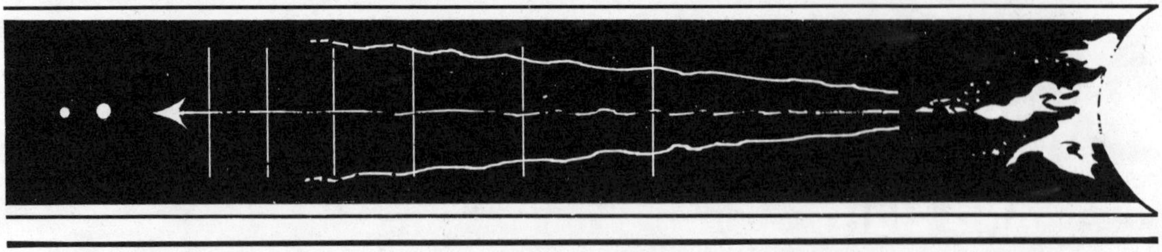

Activistatom • **DIVISION OF TECHNORESEARCH CORPORATION**

ACTIVISTATOM COMPANY, • H20 PACIFICA PLACE, • TECHNOLOGY, EARTH AD2000, • (LUV) SUR-VIVE

You receive your assignment informally from the art director. A layout artist, John, is finishing the modifications on the layout for a two-color front-page design, and the back of the sheet is one-color in the same format as the previous catalog sheets. John has the copy with the layout. Kate, an illustrator, has started the line art, and you are to use the art from the spring promotion for the main illustration. Order 5M (five thousand) copies to go to sales for distribution. They will be needed as soon as possible or sooner.

It sounds simple, but a lot can happen between assignment and performance. In this example many things will happen—a number of mechanical problems that an assembly artist might have to work out, although hopefully not at the same time.

In order to get the information you need to do the format, you might go to the art file to get a proof of the previous catalog sheet, the winter quarter. Unfortunately, there aren't any left. Say you go back to the fall quarter to find a proof. While you are looking, you find the summer quarter but keep on looking and take a copy of the summer quarter just in case. When you find the fall catalog sheet and look at the format and type style, you notice that the summer sheet is in a different typeface. That presents a problem: which typeface is the right one? Go back to the spring catalog a year ago—it is the same as the summer sheet—only the fall sheet is different. Make a mental note to check this with John when you pick up the layout, take the summer and fall catalog sheets to use for mechanical specifications, and pick up the art.

The art for the main illustration is not in the files, but luckily there is a list and John has checked it out, so you can pick it up when you get the layout and copy. What about the other piece of art that Kate was doing—it is finished and has been sent out for stats—but not for the catalog sheet. Plan to order an extra repro positive to catch up with the original order. Get the P. O. (purchase-order) number, because you've never seen the art and this is the most positive way to identify it.

You see John going to his office with layout in hand. Better call the stat house soon to make sure that the second order catches up with the first and you don't get held up waiting for the stats. But what size? You can find out from the layout with the ad illustration you are supposed to use and the copy.

From John you find that the layout was done but was changed and that the revised layout has not been done. He will not be able to get to it because of a rush job, but the revisions are three circles at the side of the art about 1 inch in diameter, each with copy inside. The copy is marked up and can be sent out. The main illustration is at the printer's with another job, but John says that he worked from a position stat that is the right size. The format for the catalog sheet was changed in the fall quarter, so forget about the style on the summer sheet. Looking at the fall format, you see that the art runs from side to side at the bottom of the sheet. John has seen the illustration that Kate did, and measuring the proof of the fall catalog sheet, which you have, gives you a size of 41 picas. He gives you the unrevised layout with instructions for the position of the three circles. A swatch of the second color is attached to the layout.

Now is the time to organize the work. You have all the materials—at least they are accounted for. The most time-consuming element is the typesetting, and that can be started right away. Call the stat house and see if they can do the second order with the first. Yes, they have the art and can do it. You arrange for the messenger to pick up the P. O. from you when he delivers the stat, because your work is a different job number. Call the printer. Yes, he has the art. Tell him to plan for a 5M two-color one side, one-color second side catalog sheet, which is needed as soon as possible. You are starting the typesetting and the paste-up and will use the art he has. The printer needs time to get the paper and will reserve time on a press.

The following descriptions and keyed illustrations explain the actual art assembly.

1. Check the copy against the layout. Where is the added copy for the three circles? Panic! Luckily it is written alongside on the art layout with the specs (specifications). Add it to the copy sheet and mark the specs. John has speced the type on the copy sheet and the type for the format, fortunately. Make sure that the type specs are complete, because you will have to keep the layout for your paste-up, and the typesetter will only have the copy to work from. There is no logo, but you can pick that up from the art file. Write the P. O. for the stat that is coming and for the typesetting, package the copy with the P. O., and put it in the pickup box. Back to the art file to find two logos, one for the front of the sheet and one for the back, and you're ready to plan the paste-up.

You have to know whether you need reverse stats from the typographer, which requires you to make some plan of how to handle the art. In order to facilitate the production, say that the type will be straight up because the logo is not reversed. You can send them out together, since an extra negative is needed anyway to reverse the logo. Have the platemaker reverse the type panel, or, if there is time, you can group the type with the logo and send it all out for a reverse stat. The art is done with the front and back of the catalog sheet side by side. The two-color front side will have an overlay for the color. The type can be in position on the flat with instructions to the platemaker to reverse it. This is not the cheapest way to handle the art, but the policy is to accommodate the production schedule without reducing quality. This often means using more expensive but faster methods. The problem is the missing illustration, which is with the printer. If you had this, you could do the paste-up as camera-ready art by stating and screening the art to size and reverse-stating the type. You have to use the stat of the illustration as a stripping guide. The original type will be provided for the cameraman to shoot. This is more complicated and requires extra negatives, but you get better quality if the camerawork is done directly from the original art.

2. The format of the back of the catalog sheet has a heading treatment that allows for flexibility in the depth of the copy, but a lot depends on how deep the illustration that runs from side to side at the bottom of the sheet is. With the logo also at the bottom the back side could get crowded.

3. Get the art materials and start the paste-up. The flat, protective flap, overlay, and tissue can be prepared from the layout. Guidelines for the format can be drawn, and some of the inking can be done.

4. When the stat of Kate's illustration comes back, put it in position above the logo. The logo position is established by the format for the back side and by the layout for the front. The logos are in position, and the rule is completed at the top of the back sheet, as are the crop marks and the register marks for the overlay. The previous fall-quarter sheet can serve as a guide for the spacing when you position the stated art. When it is in position, you are ready to do the ink border. This paste-up is literally done from the ground up, even if it is the reverse of the usual order.

5. Do the inkwork for this art on the stat—it may have to be moved to fit the type. Even if the type counts out right, you never know. What's this? The stat house sized to the art and not to the crop mark. This presents a problem. The art is too big, it won't fit the measure, and there is no way to fake it. Find out if there is time for another stat. There isn't time! John solves the problem. Slit the art at the right side between the elements and close it up to the correct width. Cut the three elements apart at the top and respace.

6. The rule for the bottom piece of the art can be inked on the stat, but the top piece will have joins when it is cut that will be very difficult to ink over—trim the art closer and do the inking on the flat. Ink the rules for the bottom piece of art, trim the art, and position everything. Draw the rules around the top piece of art in nonreproducing blue—wait until the type is in for the final inking. More can be done on the front side while you wait.

5-15.

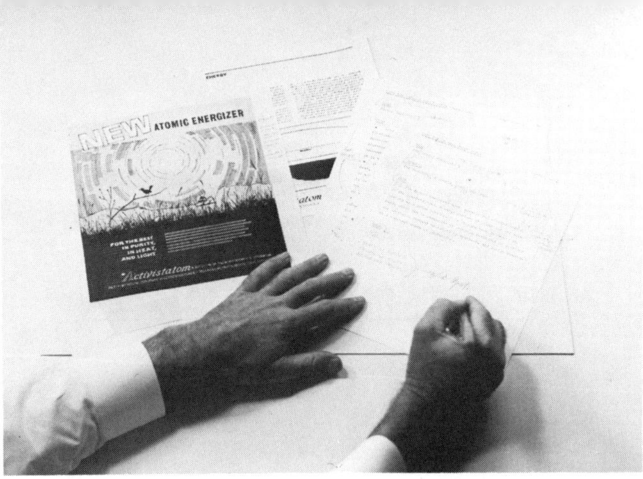

1. After you assemble the materials, check the type specs. Three subheads need to be added.

2. Start the paste-up—draw the guidelines and do some of the inking. The end of the rule can be scraped square with a clay-coated mounting board.

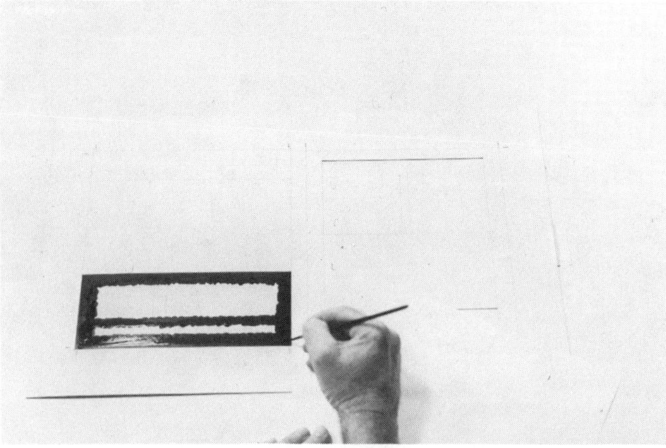

3. Ink in the reverse panel under the illustration so that you are ready to paste down the type. A piece of paper under your hand will keep the flat clean.

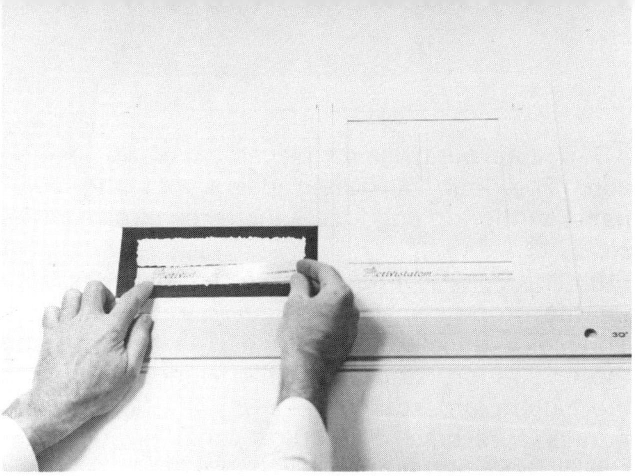

4. The design may get crowded when the logo is added. There is no layout for the back page—only a format.

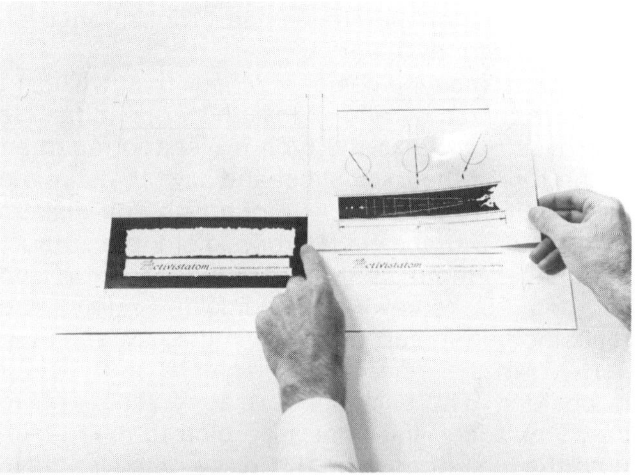

5. The stat house sized to the art and not to the crop mark, so the art is too large. Slit the art and close it up, but ink the rules first for the bottom piece, because they are so close to the art.

6. Put the stats in position—it looks as if the copy should fit. Close up the bottom piece of art.

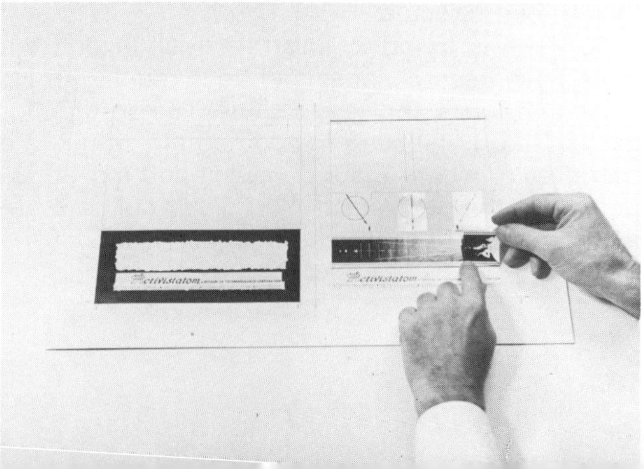

7. The layout gives the placement of the position stat. The logo is pasted down—put the color panel at the top and locate the three circles at the side.

8. The color overlay can be completed except for the dropout inside the letters in the heading. The color panel runs "behind" the heading and the illustration, stopping at the silhouette edge in the illustration. You can see why the bottom section is a reverse. The color panel prints under everything but the first word in the head, the three circles, the copy, the "A" and the small type in the logo. This makes the black appear richer. The circles drop out the illustration on the black plate and the color plate, and the inside type is black. The type and art for the circles can be positioned on top of the stat of the art.

9. When the type comes in, check the fit. The art director comes by and checks the work so far. He approves but thinks the type leaves too much air at the top of the back sheet and says to move the three units in the top piece of art up and make a tone panel with the circle dropped out to distribute the space. He marks the depth for the panel, and you begin by redrawing the guidelines in blue and repositioning the art to center in the new space.

10. Trim the type, as in chapter 4, and paste it in position. The back sheet is done according to the format, and the front according to the layout. With the type in position the inside of the letters in the first word of the heading can be dropped out of the color overlay. The edge of the overlay should center in the line.The overlay over the back sheet for the tone panel is finished, and the art is now complete, even with all the odds and ends straggling in and the deviations from the layout and format. John should see the paste-up for approval, then the art director.

11. John is trying to finish the rush layout, but there have been changes and he is redoing it. He stops and looks and says it's fine. The art director okays the art, initialing the corner, but points to a rough edge where the art was slit and moved together on the back sheet. You didn't notice it before, probably because you were so close to it.

7. Find the location of the three circles by dividing the space in this fashion: put the 0 on one line, the 6 on the other. Divide by three spaces, which puts a mark every 2 inches. This is much easier than figuring out the actual measurements.

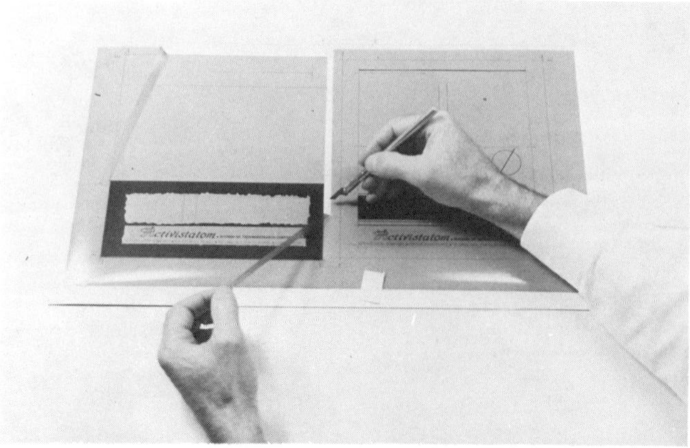

8. While you wait for the type to come back, start the color overlay and trim away the outside area. Ulano Amberlith is used in the illustration.

9. Make the art director's change, draw the rules, and mount the type. It is easier to cut out the circles separately and mount the type individually.

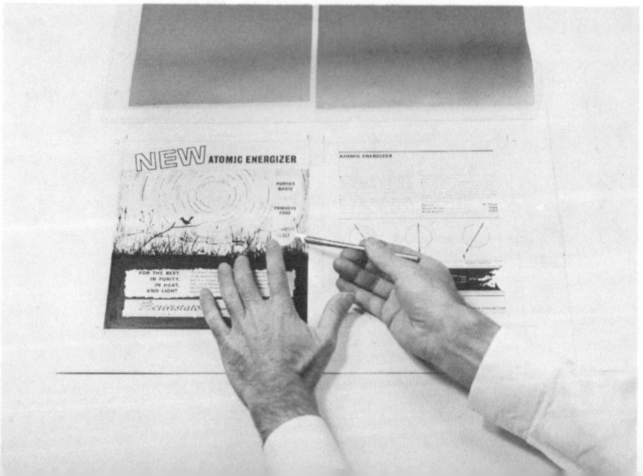

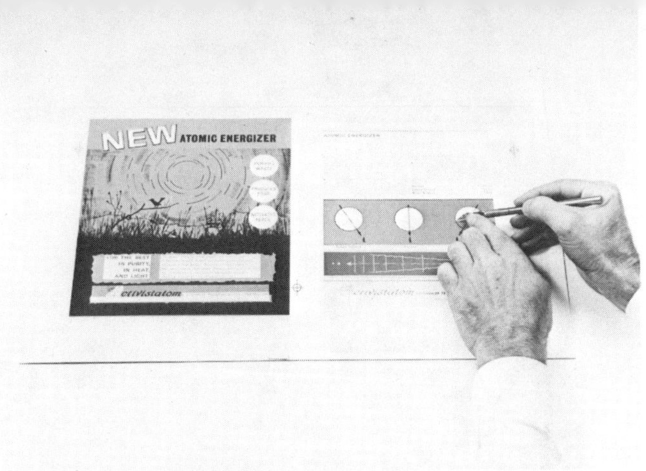

10. Finish the overlay by cutting and pulling away the areas that do not print in color and tone. Split the line in the word "NEW." The circles are a tight color register. Underlap the black halftone circle about 1/64 inch so that the black halftone closes the circle.

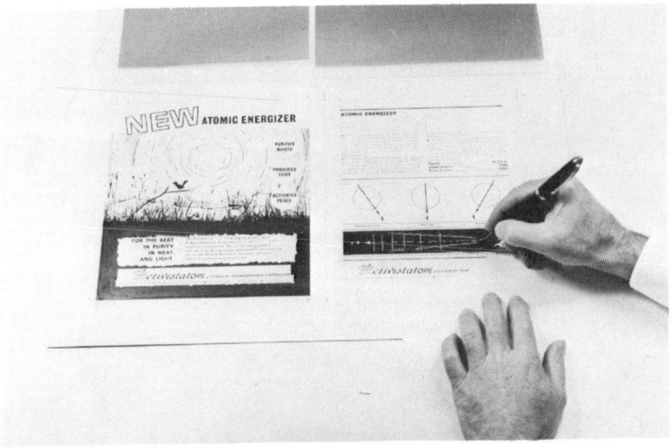

11. Ink in the cut edge in the black area of the photostat, or a thin line will pick up in the negative.

12. Sighting down the type and art is helpful in checking the alignment.

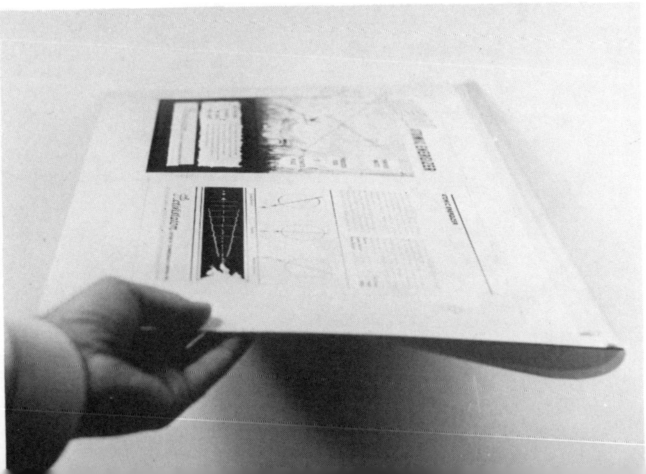

12. Small misalignments stand out when you sight down the type as if it were a rifle barrel. The art director wants the art marked up and returned to him because it must go to management for approval.

13. In looking at the art again at the drawing board, you wonder how anyone could misinterpret the requirements of the design. Closer inspection and a more objective view reveal that the overlays can both be in color. The copy on the reverse panel can be shot as is—a white panel on a black background. Part of the logo is in color. These instructions need to be explained. Using colored and black pencils or felt markers (make sure that the markers don't bleed through the tissue), render the design on the tissue overlay. It does not need to be letter-perfect, but it should show the black, tone, and color portions clearly. Attach the color swatch to the tissue. In some instances the layout can be attached to the paste-up to show the instructions, but in this case it would not include the three circles. Mark the overlays—the front as color, the back as 30% black tone. Also mark the reverse type, with arrows and brackets pointing to it.

13. Mark up the art on the tissue.

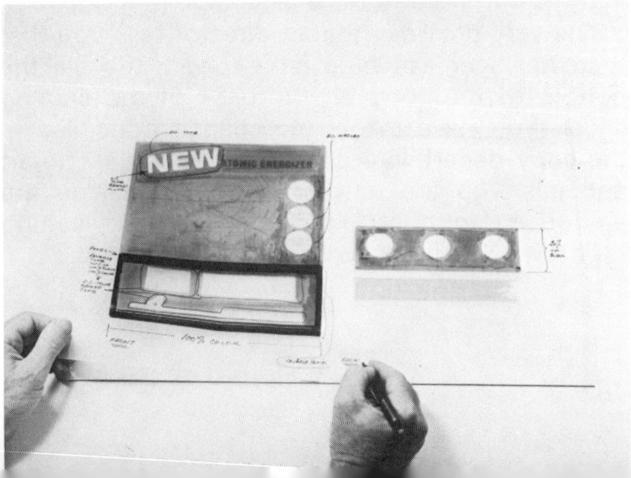

14. Note the same information on the side of the flat outside the printing area, with an arrow indicating the type—do not mark inside the printing area. This red notation is a reminder that will show on the negative but will be covered on the stripping flat. The comprehensive instructions are on the tissue overlay. Use a colored pencil or pen to indicate the second color on the tissue overlay. Draw lines and brackets to the colored type. Draw a red line across the position stat of the main art on the flat, but do not touch the edge of the solid panel below, because the platemaker needs this edge to paint up the joining of the panel to the art. Write over the stat "for position only." Outside the printing area and next to the illustration write "you have art." Explain what job it is with and whom to contact to locate the art. If the art had been sent with the paste-up as is usually the case, you would mark "art herewith." On this particular piece of art it is a good idea to make a note on the tissue overlay, with an arrow pointing it out, that the illustration also drops out in the corner of the letter that notches into it. With two or three people stripping the negatives this would be easy to miss.

15 and 16. Deliver the art to the art director. He marks the globe outside the "A" in the logo to drop out of the color plate. Take the art back to the board, correct the color overlay, and return it to him with the P. O. for the printer. In the informal atmosphere of this art department he will give it to the printer after management has approved. In other art departments there are more formal channels, and other approvals are needed before the art is sent to the printer. Here the purchase order serves as the main check and record of activity. Mark on your copy the date and time that the art left you and went to the art director.

The very next day the art director tells you that there has been a change in the specifications at the bottom of the copy on the back of the catalog sheet. You have to get the change from Max in the copy department and follow through. The art and paste-up are now at the printer's. You are authorized to print immediately without waiting for brown-line prints to check the negatives.

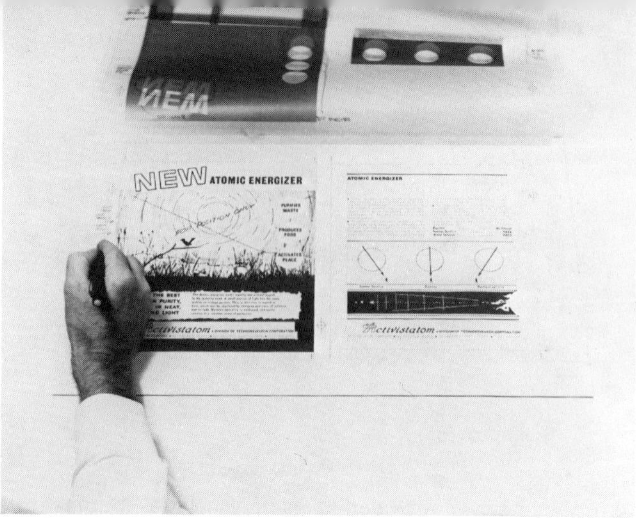

14. Mark up the art on the flat also to remove any ambiguities.

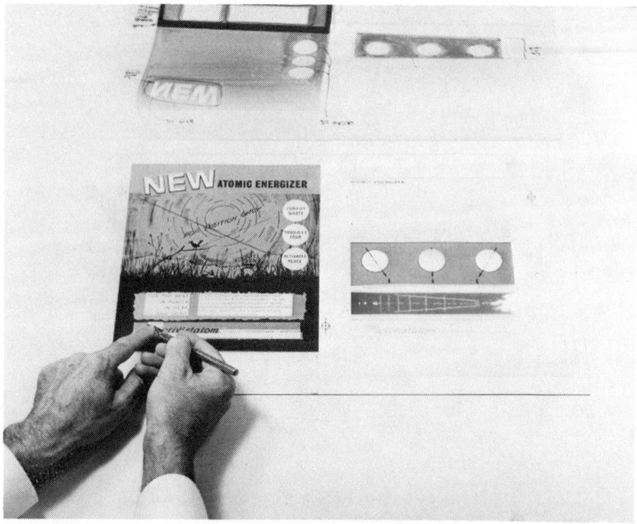

15. Correct the color in the globe outside the "A" in the logo.

16. Complete the art, check, and flap.

Call the printer to advise him that a change is coming. He catches the art in the stripping department and tells you that he has flagged it to be held for the specification change to come. Go find Max and the new copy. Max is on his way looking for you. He explains that "free" is now more expensive and that this change came through from management this morning, but the catalog sheets are still needed for a trade show at Timbuctoo Island Paradise.

Call the typesetter, explain the problem, and give him the copy over the phone line for line. You have the old copy and markup, the layout, and the new copy in front of you. Both of you are working on the same proof, which simplifies what can be a hazardous undertaking. The typesetter will set the type immediately, send repro copy to the printer, and bring a checking proof to you. Get back to the printer and tell him what is happening. There is a problem: he tells you that the last flight to Timbuctoo Island Paradise is at 12 A.M. They will have to use two presses and go into overtime in order to make it, and even this is chancy if the type change is delayed. You can't authorize this, but you will check and get back to him immediately. You look for the art director, but he is out to lunch at a sales meeting. You look for Max—he has made the mistake of having his lunch in his office. Max understands the problem, knows whom to contact, and gets a call through to the sales meeting. You get the go-ahead and a contingency that, if the catalog sheets miss the 12 A.M. flight, they are to be sent to Debarkation Coast and ferried in the 6 A.M. supply speedboat.

Hoping that you won't have to spend the evening calling Debarkation Coast, you return to your desk to call the printer. The proof for the corrections is on your desk. You call the printer and find out that he has the type, the plates have been made, and one press is in makeready. You authorize the second press and the overtime, and give him the contingency shipping instructions. The printer knows the 6 A.M. supply-speedboat route but thinks they can make the plane because of very fast service on the type alterations. The art director comes back from the sales meeting and looks at the type proof. Fortunately it's okay, but many things can go wrong doing type over the phone. You report that the correction has been taken care of and that you had to authorize overtime. He tells you to note it on the P. O. and walks away whistling. You really should call the typesetter and thank him.

5-16. The clip sheets on the following pages are duplicated at the end of the book.

CLIP AND USE THIS PAGE FOR MATERIAL TO DO PASTE-UP

Activistatom. DIVISION OF TECHNORESEARCH CORPORATION

ACTIVISTATOM COMPANY, • H20 PACIFICA PLACE, • TECHNOLOGY, EARTH AD2000, • (LUV) SUR-VIVE

Activistatom. DIVISION OF TECHNORESEARCH CORPORATION

ACTIVISTATOM COMPANY, • H20 PACIFICA PLACE, • TECHNOLOGY, EARTH AD2000, • (LUV) SUR-VIVE

ATOMIC ENERGIZER

PURIFIES WASTE
PRODUCES FOOD
ACTIVATES PEACE

FOR THE BEST
IN PURITY,
IN HEAT,
AND LIGHT

The Atomic Energizer works equally and without regard to the quantity used. A small portion of light has the same quality as a large portion. This is also true in regard to heat, which may be regulated by changing positions of solstice and latitude. Dormant mentality is awakened, and purity remains at a constant state of perfection.

ATOMIC ENERGIZER

• Always the best quality, the Atomic Energizer works tirelessly, soundlessly, and with no discernable trace of deterioration. It is literally a product with a lifetime guarantee. Truly, no one has ever made a claim against the warranty, and no other manufacturer can make this statement. It is the finest of products.

• This product works individually as well as collectively. Heat, light, energy can be obtained in the most minute amounts, as well as being available in the greatest quantities. Supply is immediate. While programmed in a cyclic delivery pattern for seasonal convenience, the Atomic Energizer actually operates at a constant rate.

• The energy service is available to each user in any amount. No consideration is made for volume of use, since this does not diminish the product, and metering is not required. Users report a slight change of skin color at first, but this is considered by many as the mark of a satisfied user, a mark of distinction.

Equinox . . . . No Charge
Summer Solstice . . . . FREE
Winter Solstice . . . . FREE

Summer Solstice **Equinox** **Northern Solstice**

NEW

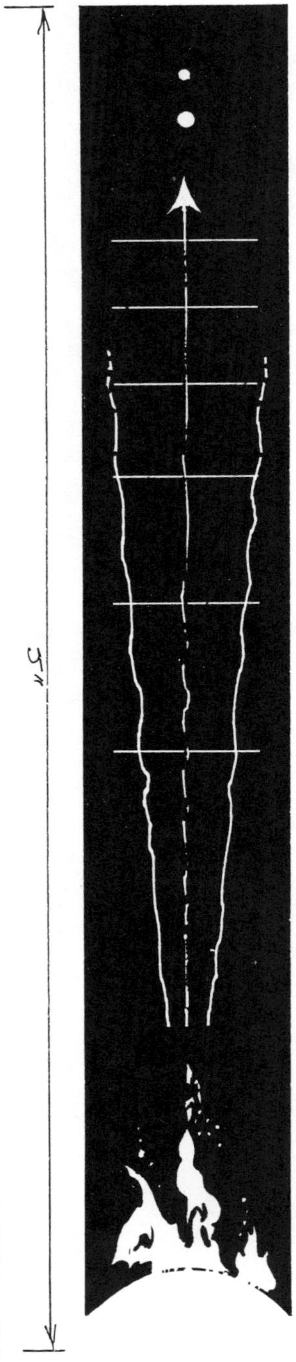

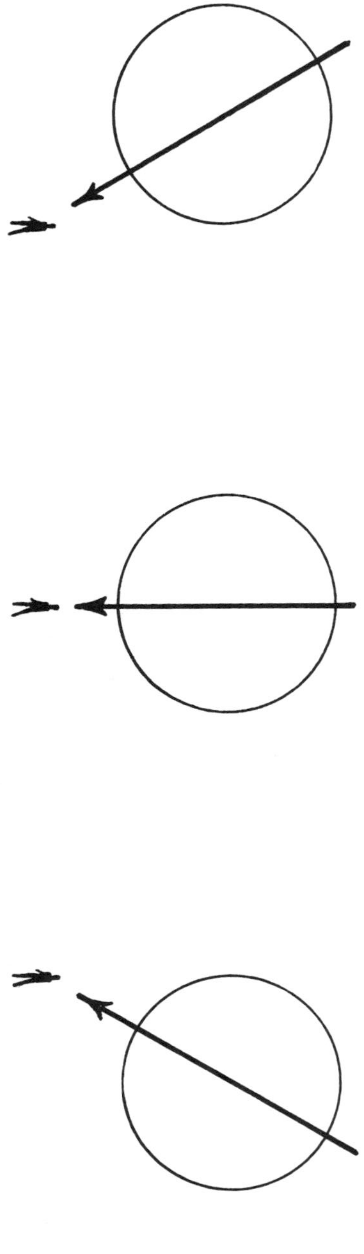

PASTE-UP

6-1. This print shop, circa 1890, points up the changes that have occurred in the graphic industry. The Platen jobbing press, made in 1874, was manually operated—note the treadle. The press had a special design feature—a pause in its motion so that it could be fed without actually stopping it—the printer fed with his right hand, removed with his left, pumped the treadle with his left foot, while standing on his right. Some wag observed that the printer had to be in pretty good shape, because his work really kept him hopping. The type cases held hand-set type. (Equipment courtesy of the Ernest A. Lindner collection.)

6. Art Preparation

Offset printing is emphasized in this book because it is gradually replacing letterpress printing. Paste-up artists work primarily for offset printing, although the basic technique is the same for both. Although the printing processes are different, the art preparation for a letterpress engraving of the art and type is identical. The major difference lies in stripping the negatives and exposing them onto the printing plate.

To make a letterpress engraving, the negatives are exposed to a prepared surface of zinc or copper and the metal is etched away, leaving only the printing surface, which is raised for inking and printing. The engraving is trimmed, excess metal is routed away, and the engraving is mounted to type height for printing. For offset printing the prepared surface of a thin aluminum-alloy plate is exposed to light passing through the negatives—the same as with an engraving—but then the plate is chemically treated without the etching process. Both processes require negatives to make the printing plate, but the offset plate is simpler because it does not have to be etched.

Still more techniques are available that further simplify the printing process. Photocomposition produces the output directly on film, which is then used to make the plate. This process does the assembly in film as stripping and not in paper as paste-up. It is similar to the technique often used for intaglio printing, in which the film stripping is done to an accurate layout or stripping guide drawn on tissue paper. One manufacturer has developed a photo technique in which a paste-up is exposed directly to a metal plate without internegatives. This process is similar to instant printing, except that the metal offset plate is good for 100,000 impressions.

New developments in the photographic processes that produce a printing surface for offset, letterpress, or intaglio continue to appear. The printing processes themselves have not changed in principle—the techniques used to put the image on the printing plate have changed. The art must still be assembled, either by paste-up or film stripping, and paste-up is preferred because it is simpler to handle a positive image on paper than a negative image on film.

Platemaking

Paste-up of material for each of the three printing processes is handled in approximately the same way. The paste-up artist works to the needs of the cameraman and the stripper, however, and should know the basics of camera reproduction and film stripping. It is not necessary for the paste-up artist to know more than the rudiments of filmwork for printing—his expertise is in paste-up—but it is helpful if he or she can visualize the film product of a paste-up. This understanding aids him in preparing the art, because he can visualize the steps that are needed to make the printing plate.

To illustrate this relationship, consider the catalog sheet in chapter 5. Little was provided for the platemaker other than the material to shoot and the positioning instructions. The type was in position, but an additional negative was needed to reverse the type image, and it in turn had to be combined with the negative of the panel that the type was positioned on. The illustration needed to be shot separately and the edge painted up to join, which required three more negatives since the original art had an overlay. The back of the catalog sheet was in better shape—it required three pieces of film but no handwork. The back needed a negative of the main flat and a negative of the overlay, which created a window for the third negative of the screen film to be mounted against. Had the art and the time been available, the sheet could have been done more simply in terms of the stripping and the number of negatives.

The cameraman and the stripper prepare the art for exposure to the printing plate. The entire process is referred to as *platemaking.* A reverse in platemaking is a clear piece of film with only the image in black. This permits light to go through to the printing plate so that the image will reproduce white against a solid field. A dropout is solid against clear film. An overprint is a second negative, in register with the first negative, that is exposed on the printing plate after the first negative. It is sometimes called a *double burn*—a process in which two negatives are exposed sequentially in the same area. A tone panel is achieved either by mounting a screen film to the area to be toned or by a double burn. An outline halftone has a solid black field around it. If a dot image remained in the background, it would have to be painted out or masked so that it would not be exposed to the printing plate. A knockout overlay is helpful for an outline halftone, because the halftone can be shot separately and followed immediately by the knockout, which is shot as line at the same reduction. The resulting negative, with a window the exact outline of the halftone, is sandwiched against the halftone to hold the edge. This allows the halftone to be given ideal exposure without having to consider whether a dot will reproduce in the dropout background.

Screened photoprints can present the paste-up artist with the same problem—an unwanted dot can appear in the background and will have to be painted out with white paint at the reproduction size. A screened photoprint made with a knockout will always have a clean background. The screened photoprint is line art and will allow dropouts or linework to be added on the paste-up flat. This can be a very useful and economical way to prepare art, since the paste-up artist is doing what the platemaker usually does, but shooting from a screened print does not obtain the same tone scale for a halftone as shooting from the original. The copy has to be shot so that the very fine dots are held when it is reshot as art. This tends to cut the tonal scale as much as 30% from what could be expected if the art were not once removed.

Although material for offset, letterpress, and intagio printing can be handled with the same technique, in some cases you may also have to prepare a stripping guide or a guide for makeup in metal.

A stripping guide can be done as a paste-up, using position stats and proofs, or as an accurate tissue drawing with the position information. Some printers who use giant intaglio rollers prefer to prepare the film themselves according to the needs of the process; others accept continuous-tone film. In either case line and halftone art is prepared together as paste-up—it is not separated because a fine screen is placed on both. Colorwork requires a stripping guide if complete separated process negatives are not supplied. Each color and each color plate must be printed separately, even

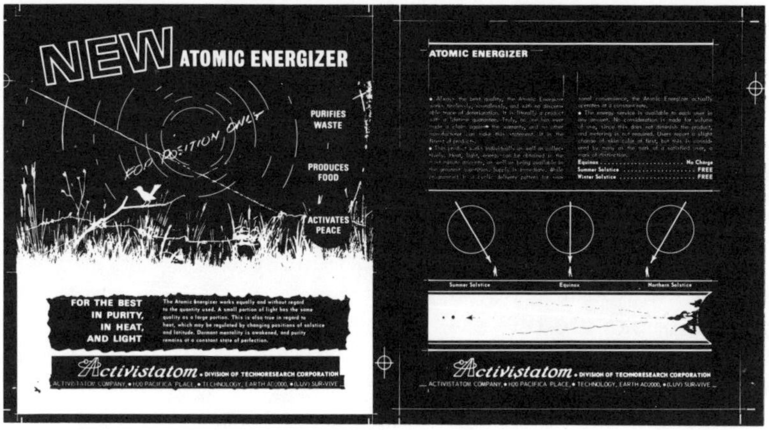

(1) NEG OF FLAT

(2) REVERSE OF TYPE

(3) NEG OF COLOR OVERLAY

(4) NEG OF TONE OVERLAY

(5) NEG OF ART

(7) NEG OF TONE

(6) NEG OF ART OVERLAY

6-2. These are the negatives that the stripper needs to make the plate for the catalog page in chapter 5. The negative of the color overlay is used to make the color plate, which prints separately—first in this case. A fatty negative can be made of the type to bring the color closer to the edge—this is done if the absence of color under the black ink is noticeable, which depends on the paper, color, and ink used.

if the art is prepared as a unit for registration.

Offset processes that produce plates without internegatives require a slight variation in technique in that the art must be complete for shooting. Filmwork for photostats and screened prints must be done prior to paste-up, and the paste-up uses line art only.

Letterpress is often used for information that is constantly updated. Telephone books are one example. While there are techniques in which information is stored in computer memory and arranged automatically for printout, the economics are such that letterpress is the better choice in some situations because of the ease with which new material can be respaced. Art material such as an ad can be done as paste-up, with a complete engraving for lockup, or else the art alone can be engraved, with the ad done as makeup. If the work is done as makeup, a layout tissue or a pasted-up dummy is necessary as a guide. Since the dummy is used only as a guide, it need not be done as carefully as work for reproduction. The stoneman who does the makeup in metal and lead spacing from the guide will see to the exact alignment and spacing for reproduction.

6-3. A loss in tone scale occurs with a halftone as compared to the original. The white was originally pure paper white; it now has a small background dot in it. The black was solid black; it now has a slight gray cast. Dropping out the background makes the white a pure paper white. You can also use a screened photoprint to make the whites white and the blacks black by applying white paint and black ink directly to the print. Shooting from an unretouched print, however, narrows the tone scale about 30%. Compare the loss of tone value in this art to the reshot halftones shown in figure 4-17.

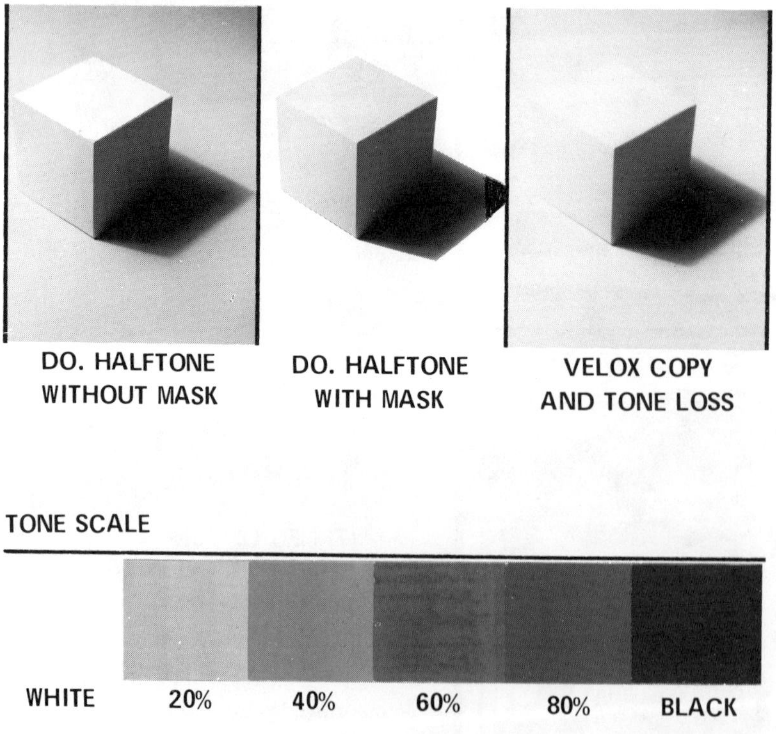

Binding

There is an enormous variety of sizes and techniques available for folding and binding printed material. The work is specialized, and smaller printers often send material out to be bound. Large publication printers have a bindery on the premises, sometimes with different techniques and specifications available.

Binding is part of the design function: the designer needs the folding, collating, and binding restrictions before starting the design, because this information sets size and shape limitations on the design. In practice, the completed design or format is presented to the paste-up artist for production—he is not involved with the initial decisions. Many factors play a part in assembling the physical printing package. Quantity and size of paper, its availability, press size, color, bindery capability, shipping—all must be weighed between cost and quality to arrive at a final package.

In general, printed work is done in multiples, often of sixteen, on as large a sheet size as is economical in paper costs to reduce press time and to arrange signatures for binding by folding rather than collating. In the work-and-turn method, for example, the complete job, both front and back, is run on one side of half the amount needed; the work is turned so that fronts and backs match, and the other side of the sheet is printed to complete the run. This technique uses one instead of two printing plates and simplifies press makeready. Similarly, brochure pages are arranged so that a number of folds puts them into proper order, and the edges are trimmed after binding to make separate leaves or pages.

The common practice in paste-up is to do pages in reading order because the continuity is visible and two-page spreads rather than single-page units can be designed. The cameraman shoots the spreads; the negatives are cut; and the stripper puts them on flats in proper press imposition for the printing plate. Paste-up can be done to press imposition, but it makes exact spread alignment nearly impossible and continuity difficult to follow. Large drawing boards are needed, and T-squares and triangles must hold the alignment. Any misalignment means that the work must be cut apart and realigned by the stripper, which eliminates the savings gained by pasting up to press imposition. If there are also color overlays to register, the danger of shrinkage or expansion in the overlay arises. The camera operator must shoot and process the film for a paste-up done to imposition as one unit, and this can present problems with both exposure and handling, particularly for a unit larger than about 24 or 26 inches.

If you are involved with binding and press imposition, talk to the printer and the bindry and find out what they like to work to and what they offer. Preliminary work with the specialists can save time and money and make the work easier for all concerned.

6-4.

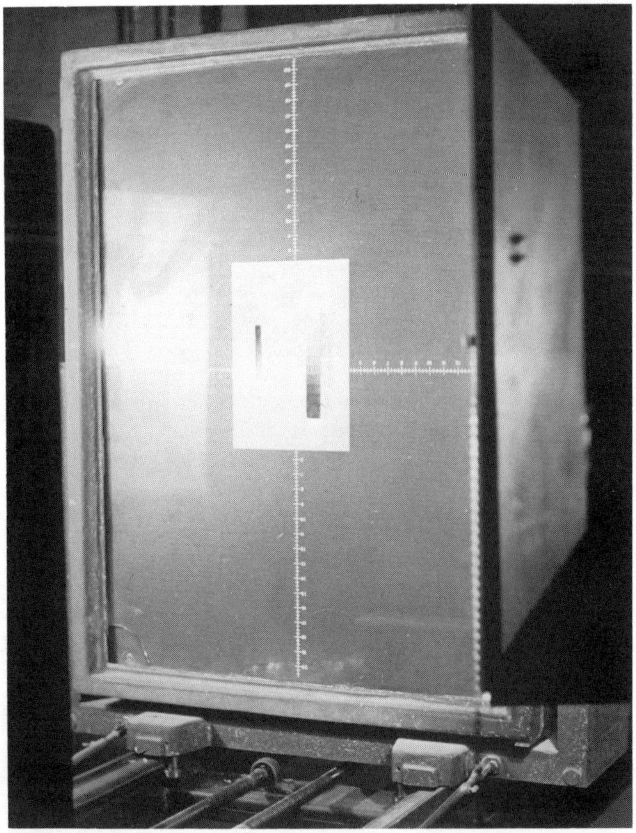

1. When the art arrives at the lithographer's, it goes to the giant copy camera. The copyboard on this particular camera measures 40 inches lengthwise.

2. The photo stripper's light table and equipment resemble the artist's drawing board and taboret.

3. The photo stripper tapes the negatives onto goldenrod paper—a heavy orange-colored paper that blocks light. Pins are mounted to register the different levels of negatives.

4. The metal dowel pin and tab shown here make an excellent register. Notice the window for a halftone in the negative on the light table. The stripper works with the mirror image of the copy: the negative is emulsion side up.

5. A reverse-type negative is taped into position. Notice the drafting instruments and the scale. This work is as precise as paste-up.

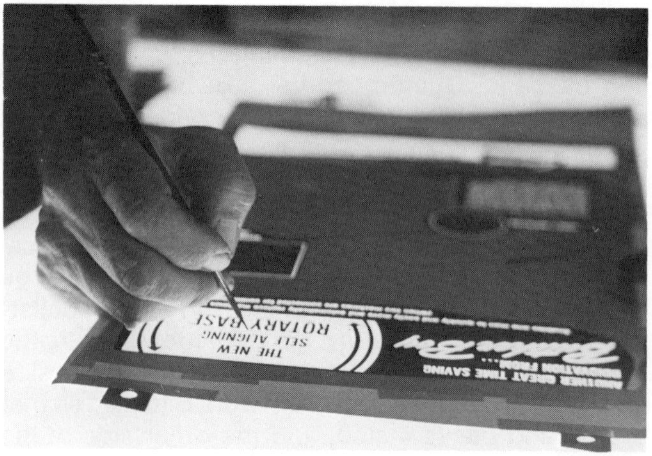

6. With the stripping complete, the flat is turned emulsion side down (right reading), and any additional pinholes in the emulsion that would admit unwanted light are opaqued out.

7. New levels of goldenrod are mounted, and additional negatives are stripped.

8. A sensitized lithography plate is placed for exposure. The dowel register pins permit sequential exposure of the negatives that in combination make up the complete printing image on the plate. The sparkling-clean glass catches the reflection of a fluorescent light.

9. The plate is exposed to powerful arc lamps. It is held in a vacuum frame that keeps the negative emulsion close to the litho plate.

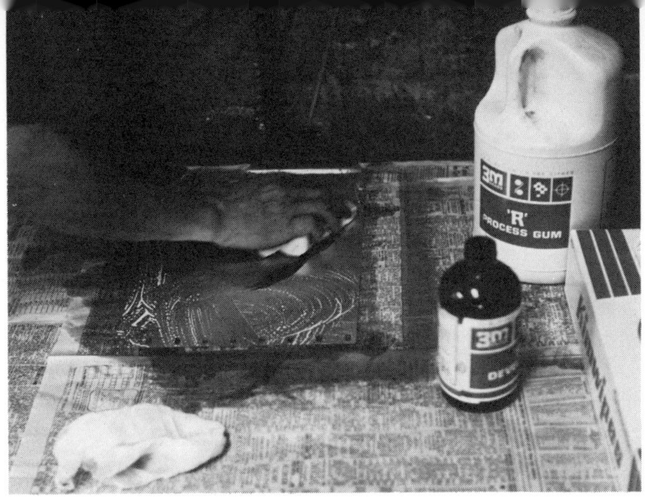

10. Gum is wiped over the exposed plate to maintain water receptivity.

11. The developer is wiped over the surface, which brings up the image area to be inked. The nonimage water-receptive areas are ink-repellent; the image areas are ink-receptive and water-repellent. The fundamental principle of lithography is that oil and water do not mix.

12. The plate is now mounted to the press for the press run.

13. Compared to the 1874 hand-operated press shown in figure 6-1, these presses look huge and complicated, and some are automated. Automated presses operate from a console that looks somewhat like a computer console. The atmosphere is charged because these presses run fast, and when something happens, it happens fast.

14. In either a job shop or a giant printing plant, paper is part of the atmosphere.

15. Compare this paper cutter with the one used by instant printers. (All lithography photographs courtesy of Collins Lithographers.)

7-1. The linecaster is the traditional way to set type, but the machine-dominated atmosphere is changing. New systems are being developed that bring an office environment to the production of graphic work.

7. Specialist Fields

Before mass-production techniques came into widespread usage, a graphic artist was master of many trades. Now there are specialists for different kinds of artwork: the theory is that a specialist can do the job better and less expensively; the rule is that you shouldn't try to do specialized art for publication unless you are a specialist; and the application is that if you need an illustration, call an illustrator.

The economic situation of many publishing and advertising companies precludes the use of specialists, and work that would be better illustrated than described in words goes into print without pictures. It is usually a question of money or time or both. The graphic industries, which require so much paste-up work, are particularly susceptible, since they often develop cold-type and in-house production systems to cut costs. Almost no publishing or advertising projects are done without a consideration of the costs involved and artwork is the first victim of budget cuts, yet art expenses make up only a small proportion of the total cost.

Economic reasons are largely responsible for the present development and sales of transfer-type and clip-art books. Clip art and transfer type, cleverly used, work almost as well as spot illustrations and type patches, and they cost a lot less because they are mass-distributed. It is the difference between carriage craftwork and mass production with options. The paste-up artist is often the person who inherits this problem, especially in small or one-person art departments, offices, and other budget situations, and clip art and transfer type are a big improvement over no art at all. Larger art departments have specialists on the staff, so the problem usually doesn't arise.

Decisions to use no artwork, to do it yourself, to use artist's aids, or to call a specialist are based on the market for the particular printed piece. For the paste-up exercise in chapter 1, for example, it is obviously better to use art and, in terms of cost and application, to do the work yourself. The purchase of clip art would be justified if there were a bigger mailing or a number of similar pieces, because a clip book usually contains a quantity of pieces. A budget for an original and specific spot drawing is justified for more specialized work or for more expensive mass markets.

These decisions are made by the art director, the designer, or, in their absence, the paste-up artist, unless they are dictated by management. There must be a balance of expenditure, application, and desired effect—the balance of cost versus quality versus speed.

7-2. Clip art for any type of job is sold on the mass market. These are only a few of the books (shown at reduced size) published by the Volk Corporation (1401 North Main Street, Pleasantville, New Jersey 08232). Volk is one of the top services and has been in business for a number of years. Clip art is an excellent value; with it the design problem becomes how to adapt existing art to the design.

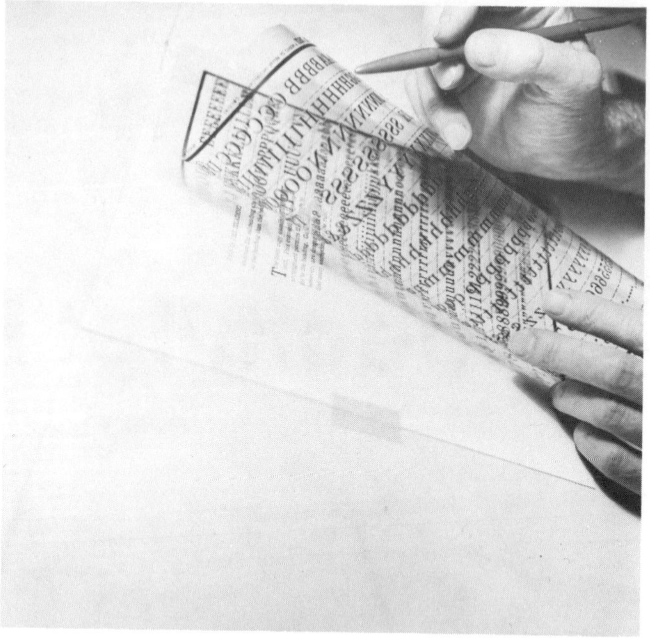

7-3. Transfer type has improved in quality and gained wide usage.

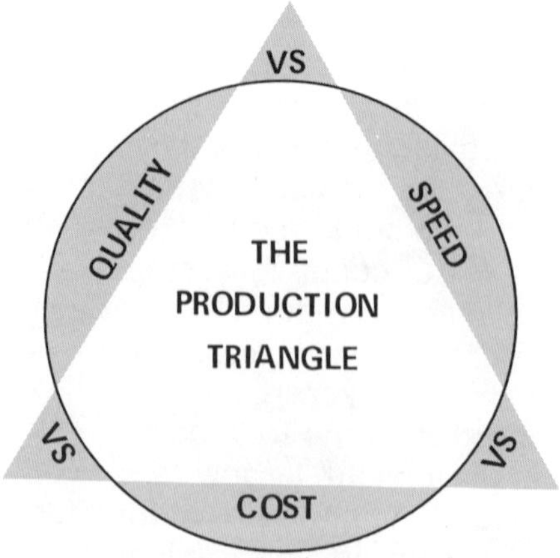

7-4. This diagram shows the three forces that must be balanced in making any design.

Graphic Specialties

Specialists develop out of a continued demand for a specific product. If there is sufficient work, an artist can specialize and promote the theory that specialization is better. In practice, a specialist can do a number of different kinds of commercial art but chooses to concentrate on a particular one because of preference or demand.

Areas of specialization that are now standard in commercial art are: art production, illustration, photography, retouching, design, lettering, and typesetting. Within each area there are more specalized categories, and among individual artists considerable differences of interpretation. In order to select a suitable artist for a particular job, it is helpful to know something about the various specializations—just remember that there are no real boundaries but rather merging areas of interest or opportunity.

Photography, for example, is often categorized as either fashion, industrial, publication, or people. According to this division you would not hire a fashion photographer to shoot an industrial subject because the information background and point of view would not be suitable—unless you are a clever art director who chooses a fashion photographer to shoot gasoline pumps as if they were fashion figures and presents an industrial subject in a way that appeals to women, thereby producing an excellent and successful advertising campaign. This shows how arbritrary these categories can be.

If the art director had used the fashion photographer with the expectation of a conventional industrial treatment, however, he would have been very disappointed. This points up the other side of the question and explains why the categories exist at all. Too often art is purchased under nebulous assumptions, with misunderstandings and disillusionment the result.

Experienced specialists learn to avoid this kind of situation because the results are unsatisfactory to everyone. An experienced buyer learns how to screen artists, how to avoid misunderstandings, and, most important, how to make the best use of the talent he or she hires. An understanding art buyer provides the right information and creates a climate that encourages the best efforts of the artist. These categories, then, are only a tool to help you understand patterns of specialization in commercial art, which can be helpful if you're in the market.

An *illustrator* is an artist who specializes in pictures. There are several kinds of illustrations and manners of approaching a subject. An illustration can be done in a painterly fashion or as a drawing that relies mainly on draftsmanship; it can be figura-

7-5. A wide variety of approaches can be found in any graphic specialty. While both of these treatments are realistic, they create completely different effects.

tive and literal or decorative and imaginative; in color or black and white; a realistic rendering or a depiction of an abstract idea. Although an illustrator could illustrate anything, he or she would not approach every subject or idea equally well and would have areas of particular preference and ability.

In a field as varied as illustration specialization seems to naturally fall into categories: general-story, decorative, rendering or technical, fashion, product, black-and-white, and spot-drawing. This list is far from complete—there are many other specialties that require a great deal of technical knowledge, such as medical or wildlife illustration. Combined with the different manners of approach previously mentioned, this list is complete enough to show that you cannot hire just any illustrator!

If you need an illustration of a structure not yet built, for instance, you should hire an architectural renderer. He is a specialist in reading blueprints and developing an accurate picture of what the structure will look like. If you asked a decorative illustrator to draw the structure, you might well be disappointed with the result, especially if you were looking for a literal rendition, and the fault would really be your own.

When you hire an illustrator, be specific about your expectations, if you have any, and provide adequate information. Don't expect an architectural renderer to illustrate a specific building without providing the floor plan and elevations. If you do need an illustration from verbal description alone, develop a sketch or series of sketches to make sure you agree on the specifications. This holds true for the other categories of illustration—fashion illustrators, for example, usually specialize in men's or women's clothing, but not both.

An illustrator will show samples of his or her work if you request them, and you should be able to figure out his manner of approach to a subject and his areas of preference. The more services you require beyond the actual illustration, of course, the higher the price will be. In situations in which specifics are required, this understanding is necessary and determines the success or failure of the work. Unless the illustrator is very high-priced and overworked, he will usually be more than happy to work with you, even to venture into areas outside his specialty, if you understand the problem and provide the opportunity.

A *photographer* is an illustrator with a camera. Because of their supposed literalness photographs are usually preferred over illustrations for products and illustrative situations that profit from an illusion of reality. Once you realize what photography can do to interpret a subject, it becomes obvious that a photograph expresses a point of view developed by an artist. Setting, lighting, and manipulation during development allow a great variety of photographic interpretation. Add to this composition, inclusion of associative material, multiple exposure, and alteration of the image during printing, and you can do just about anything with photography—what is more, you can make it seem real.

All this camera magic presents the art buyer with the same problems in buying photography as in buying illustrations, and the techniques for coping with them are the same. Provide specific information, understand the photographer's work and approach, and explain the effect you want clearly. Print-photography categories—retail, industrial, publication, and people—are more general than those for illustration. These categories are helpful, but the selection of a photographer must rely heavily on individual technique and approach.

Photographers are sought for their interpretation and understanding of particular subjects and moods. If you hired a dozen professional photographers to shoot starfish, for example, you would undoubtedly be amazed at the difference in their conception of what a starfish looks like. The objective in buying photography is to get the right approach for the job.

Sometimes you don't like the photograph you commissioned. The situation may be nearly impossible—a photograph of a building that you see as a building but the camera sees complete with signs, fireplug, telephone poles, litter, and an adjoining building. Even the best photographers must deal with this kind of a situation by *retouching*.

A famous advertising man once said, "Don't sell the steak—sell the sizzle." The retouch artist's job is to sell the sizzle. He removes offensive realities such as fireplugs and telephone poles from the print and heightens the shine. It really is a venture into surrealism—nothing looks the way it is presented, but no one would like it if it wasn't presented that way.

A retouch artist is a rendering artist and a specialist with the airbrush. With great skill and attention to detail he paints the desired image over the literal image and does it so subtly that the demarcation between the painting and the photograph is invisible. This technique is more mechanical than illustrative and borrows from the photographer's interpretation, but there is still a considerable amount of artistic ability required.

General specialist categories are: food and cosmetics, for color retouching; hard-line and soft-line merchandise, for black-and-white retouching. Considerable knowledge is required about which points to emphasize and which to play down. The art buyer should make all of this clear, preferably with examples. Subjective ideas are very hard to deal with—some people like the sizzle rare, some well done, others at all points in between. The retouch artist has to know what the dream is to visualize it.

Of all artists the *designer* deals most closely with management, because management initiates design and formulates the problem in the abstract to the designer. It can be more than a problem, especially if management is not clear about the concept and continues to develop it after the designer is already in the picture. For the designer to function at his best, the buyer must at some point make a firm decision as to what the objective is.

For the best results the designer should sit in on the concept-formulation stage, because he can often point out advantages and disadvantages to different courses of action from his point of view. Once the program is formulated, the designer initiates his contribution. He develops the program, either personally or with other specialists. He brings the defined purpose to fruition according to the prescribed requirements. There can be considerable consultation with management as to the relative value and priority of different elements in order to make final design decisions. Explicit concept formulation is vital in that it structures the process and outcome of the design—you do not get pine trees from acorns!

Because different background knowledge and information are required for different jobs, designers too are categorized by interest, preference, and specialization. The major division is between three-dimensional and graphic designers. Unless you are dealing with package design, you will almost always need a graphic specialist. Within graphic design there are editorial and advertising specialists; the first requires expertise in a subject, and the second expertise in salesmanship as related to a subject. Design approach can range from comprehensive subjective development of the public image to specific objective design of a particular subject area, such as a periodical. The design function is not nearly as abstract as it may seem: it follows definite logical patterns, structures, and reasoning processes to its conclusion.

The buyer of graphic design should make a selection according to past performance, background, specialty, and artistic ability. Samples can be helpful, especially if the relationship between management and the designer is known. All artists are in part designers—a graphic designer is simply a specialist in this area.

A graphic designer is often a *lettering* designer as well, specializing in type design and lettering. The designer may be involved with lettering, paste-up, and production in order to coordinate the design.

Southern California Edison Company

7-6. This logo, designed by Jerry Gould & Associates, exhibits the relationship between design and lettering as specialist functions. Market research by Southern California Edison Company has revealed that most people read the letters before the graphic image of a cord and plug. This design achieves an ideal balance between the letters as symbols and their pictorial meaning.

A lettering designer is often hired to do logos and letterheads. These designs are primarily involved with typography, and a lettering artist specializes in the design and rendering of letterforms, alphabets, and words. Whether a design involves creating a new alphabet or assembling transfer type, it always profits noticeably from the background and expertise of a lettering artist.

The present demand for unique alphabets has brought the lettering artist into the field of photolettering. The ability to use a relatively inexpensive font negative at any size and with any spacing has reduced the demand for handlettering as artwork to some extent, and there are also quality hand-lettered transfer-type alphabets that can be spaced by hand. Even with all this variety, there is still a demand for lettering artists—only they can create word designs and art-nouveau-type designs.

The art buyer can judge a designer's preferences and talent from samples. Few areas in art are as well defined as the alphabet symbols, yet there is an incredible and growing diversity of letterforms.

7-7. Although there is an increasing diversity of letterforms on the market, the lettering artist is still in demand for art-nouveau and dada designs.

A *typographer* can be an artist and specialist in the composition and selection of type, while a *typesetter* may simply set type mechanically, but these terms are relative—some typesetters are designers. Photolettering and photocomposition present still more scope for design, because these processes permit unlimited spacing arrangements and more unique treatments than ever before.

Computerized typesetting adds still another dimension to the storage and handling of typography. The typesetter can become a graphics consultant who provides the means for developing a graphic system, a system of production that influences the graphic form through automated typesetting.

Increasing specialization at first seemed to mean that the typesetter only set type, unlike the earlier typographer, who was also a designer and printer. Because of the need to coordinate the graphic-production system with the new methods of typesetting, however, the typographer appears to be regaining his original position, although in a new dimension.

The typography buyer has a considerable array of quality levels and techniques from which to choose, ranging from the composer to computer-driven photocomposition fed by optical character recognition (OCR). The manner of approach can range from handcraft that emulates the technique and appearance of historical work to automation that looks like nothing you have ever seen before.

If you want typesetting along traditional lines, samples help you make the right choice. If you are using the new techniques, however, you are basically dealing with a designer: the concept has to be developed, and the design of the production system worked out. In the nongraphic terminology of automation this translates into developing the software and program capabilities for the system. Whatever the terminology—each specialist area seems to develop its own—it is still a graphic-production problem and can be dealt with as such.

Typesetting categories are the traditional newspapers, publishing, and advertising, with newspapers leading the way in the development of automated techniques. Typographers specialize in either publishing or advertising.

7-8. This type lockup and its image show how graphics are produced with a letterpress. This is the system that paste-up replaces: positioning elements in paper is easier than positioning them in metal.

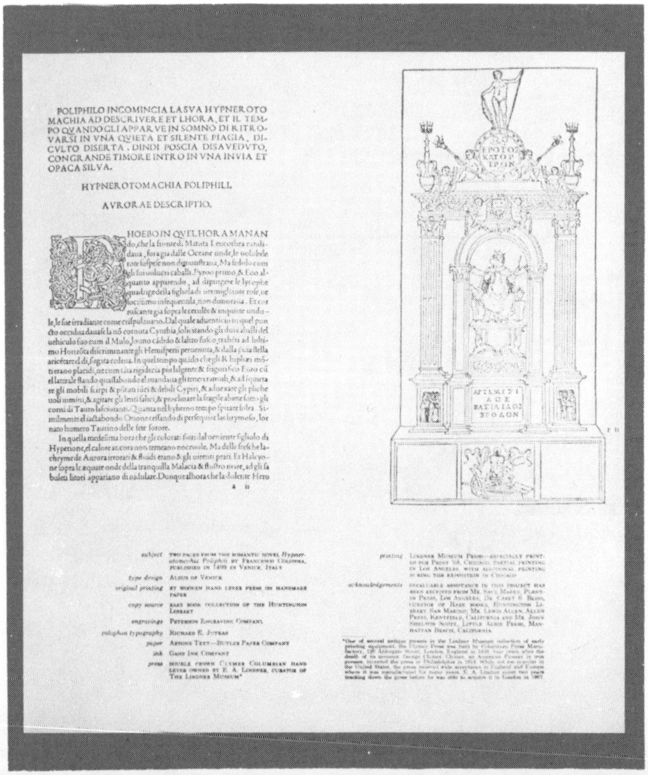

POLIPHILO INCOMINCIA LA SVA HYPNEROTOMACHIA AD DESCRIVERE ET LHORA, ET IL TEMPO QVANDO GLI APPARVE IN SOMNO DI RITROVARSI IN VNA QVIETA ET SILENTE PIAGIA, DI CVLTO DISERTA. DINDI POSCIA DISAVEDVTO, CON GRANDE TIMORE INTRO IN VNA INVIA ET OPACA SILVA.

HYPNEROTOMACHIA POLIPHILI.

AVRORAE DESCRIPTIO.

PHOEBO IN QVEL HORA MANAN

subject — TWO PAGES FROM THE ROMANTIC NOVEL *Hypnerotomachia Poliphili* BY FRANCESCO COLONNA, PUBLISHED IN 1499 IN VENICE, ITALY

type design — ALDUS OF VENICE

copy source — RARE BOOK COLLECTION OF THE HUNTINGTON LIBRARY

press — DOUBLE CROWN CLYMER COLUMBIAN HAND LEVER OWNED BY E. A. LINDNER, CURATOR OF THE LINDNER MUSEUM*

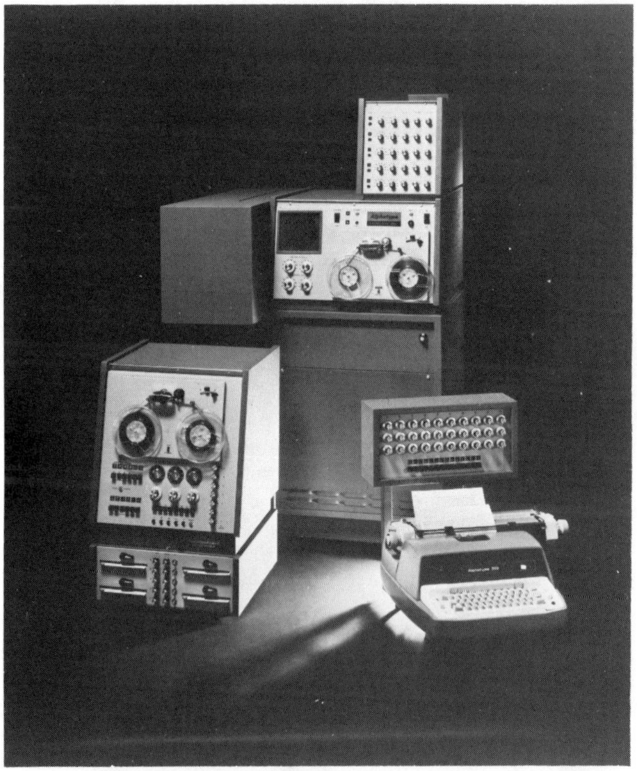

7-9. The Alphatype photocomposition equipment produces the type image on paper or film. It can be used in place of the linecaster shown at the beginning of this chapter. Underlying the design of this equipment is an entire system of concepts that utilize electronic computer technology.

Specialized Production Systems

Production systems influence specialization because they prescribe the areas in which specialization can be practiced. Conventional production systems rely on a pool of suppliers; the art department generates the system. The use of cold type can change the system, increasing the amount of paste-up in particular. Chapter 1 illustrates the trend toward an entirely new art-production system in general businesses, which in effect brings the art department into the office.

Computerized cold-type photocomposition makes possible many more and vastly different art-production systems, because the computer can store information and retrieve it on demand as well as handle output automatically. More art-production systems will inevitably be created to make use of this feature. The graphic industries and office art departments will eventually have entirely new production systems that are specialized for entirely new graphic purposes. Add to this the potential of recent equipment—COM (computer on microfilm), which coordinates computer and film; OCR (optical character recognition), which permits automatic input to electronic computer memory; photocomposition and the CRT (cathode ray tube), which allows output of typeset and graphics—and the stage is set for many yet to be defined fields of specialization.

Typography is another unique specialization for the paste-up artist. Paste-up has always been involved with typography, but only at the stage of assembling the material. This new specialization deals with preparation of the material prior to assembly. Computer typographers could certainly use people who can handle paste-up and keyboard input for typesetting, and a combination of these skills will be the format for a new job classification. Up till now new personnel usually had to be trained by apprenticing.

7-10. Computer technology makes this system possible. It is essentially a computer-centered system with a variety of input-output devices. Data, the computer term for copy, is entered into the computer memory, where it is stored, retrieved for various manipulations, then stored again for output as photocomposition. Newspapers have led the way in developing these systems because of their need for speed. These systems are fast.

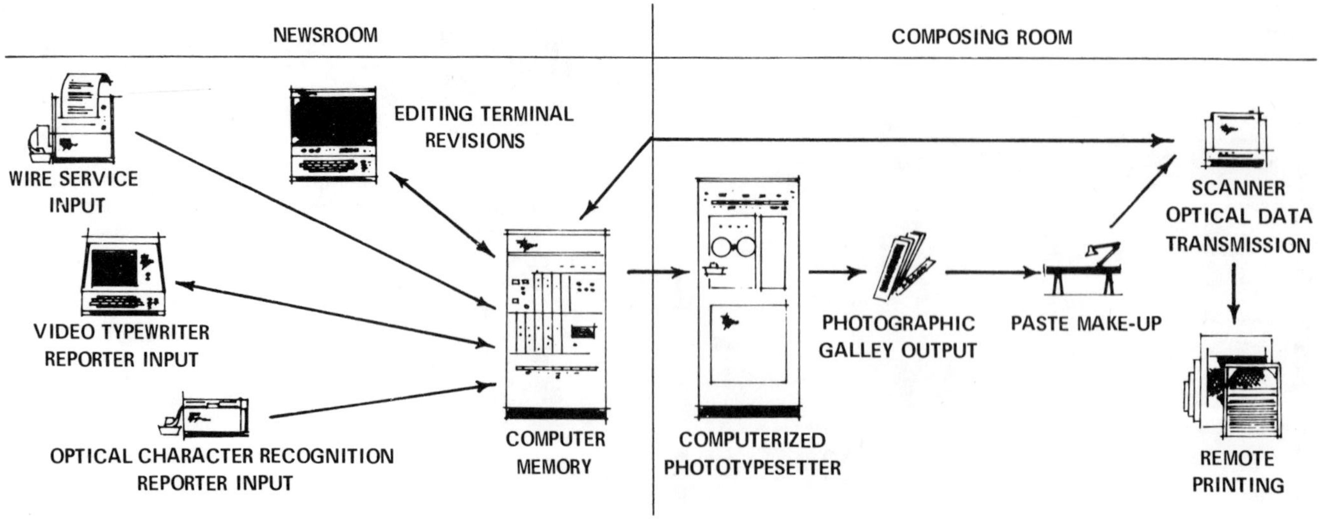

There are still other new techniques, one of which involves the use of a video terminal that permits electronic editing and formatted output for typesetting. The majority of these new techniques utilize paste-up to some extent, at least an understanding of the requirements. Paste-up is still one of the best places to begin learning about graphics. It is in a central position, in association with all the other elements of graphic production.

Production systems have always changed along with technological development. There will be considerably more change, and very likely more rapid change than heretofore. Paste-up—handling the image on paper—is basic to all of these developments and to all of the elements of graphic production.

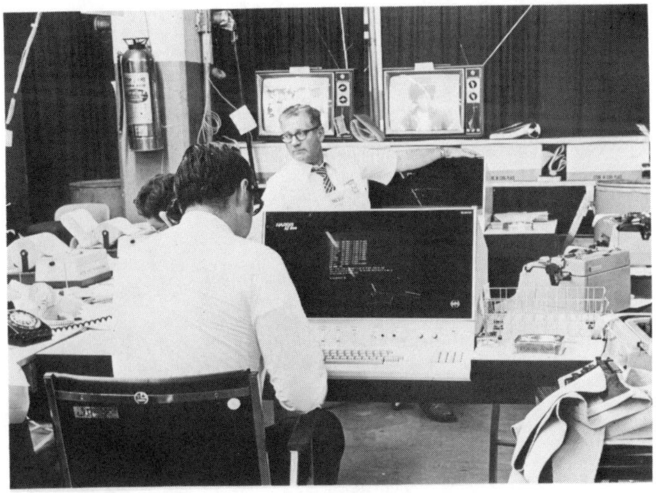

7-11. This "graphic system," used by United Press International, was developed by Harris-Intertype Corporation. The remote installation has an interstate link to UPI's central computer in Manhattan, from which stories flow to news outlets around the world. Newspapers can channel the stories directly into computer storage for editing and output them at speeds up to 6,000 characters per second, making newspaper publishing as fast as broadcast journalism. The difference, of course, is that you can hold a newspaper in your hand for continuous viewing. Picture copy can also be handled in this way, as can platemaking for remote printing. (Photograph courtesy of Harris-Intertype Corporation and United Press International.)

Specialization in Paste-up

Art production is an area of specialization that stems from paste-up, especially with the new developments in the graphic industries. Good paste-up requires aesthetic sense, design ability, and production know-how. Art direction is another specialty: it requires well-developed design sense, knowledge of art production, and ability to work with management. A paste-up artist naturally develops in this direction, unless you choose to concentrate on production management instead. The determining factor is whether you prefer art and have artistic skills, or incline toward business management.

Paste-up can also lead to specialization in one of the supplier roles such as illustration, photography, retouching, design, or lettering. These jobs require special training and business expertise in addition to production know-how. To be a commercial illustrator, for example, you have to not only develop a portfolio of work that demonstrates your abilities, preferences, and style but also handle the business with prospective employers. This is easier when you have a few real jobs under your belt.

Art specialties are also easier to develop with schooling and training, but this is not to say that you cannot do it on your own. Many famous artists have started in just this way. Hard work, perseverance, and thought directed toward a purpose are essential even with the best training. Practice and an intelligent approach can help you accomplish your goal. The graphic industries continually present new opportunities.

Paste-up can be much more than merely mechanical art production. What *you* do with it depends on your preferences and goals and on your development of artistic ability. The fact that paste-up is essential to current graphic techniques is only a clue to what its role will be in the near future. At the very least there is an increasing demand for paste-up in the present.

Index

Duplicate Clip Sheets

The clip sheets on the following pages are duplicates of the do-it-yourself exercises in chapters 1, 3, 4, and 5. Use them if you make a mistake, for additional practice, or to develop a different idea.

You are invited to a

CLIP IT CLEAN!

FUN, GAMES AND PRIZES

Our famous egg balancing race is a featured attraction. Henrie Souffle, last year's champion, will take on all challengers. His crown is again on the line--the finish line, that is. The entertainment director promises even more opportunities for feats of greatest skill.

WHO, WHEN AND WHERE

This gala event is for the employees of the MANUFACTURING COMPANY, INC. and their families. It takes place on the weekend of the 4th, Saturday Afternoon from 11 AM to 4 PM at the Jamboree Park. Billy the Kid & his Hoe-down-slikkers will make music and call the tunes, so dress country style.

FOOD AND REFRESHMENTS PROVIDED

Your employee card entitles you to one box-lunch for yourself and each of your guests, and all the soda pop you or anyone else can drink. Door prizes given to the 5 numbers drawn from the fish bowl.

It's time for
The Annual Company Picnic!

ALL EMPLOYEES AND THEIR FAMILIES ARE INVITED TO THE ANNUAL PICNIC

MANUFACTURING COMPANY, INC.

CLIP AND USE THIS PAGE FOR MATERIAL TO DO PASTE-UP

LAYOUT

TRAVEL...

THE SOUTH SEAS

Do It Now Travel Agency

the South Pacific offers
enchantments of balmy,
soft weather, trade winds
laden with the scent of
romantic adventure...

THRILLING
TOURS

PAGO PAGO

TYPE GALLEY

ART

TRAVEL...

PAGO PAGO

REVERSE STAT

CLIP AND USE THIS PAGE FOR MATERIAL TO DO PASTE-UP

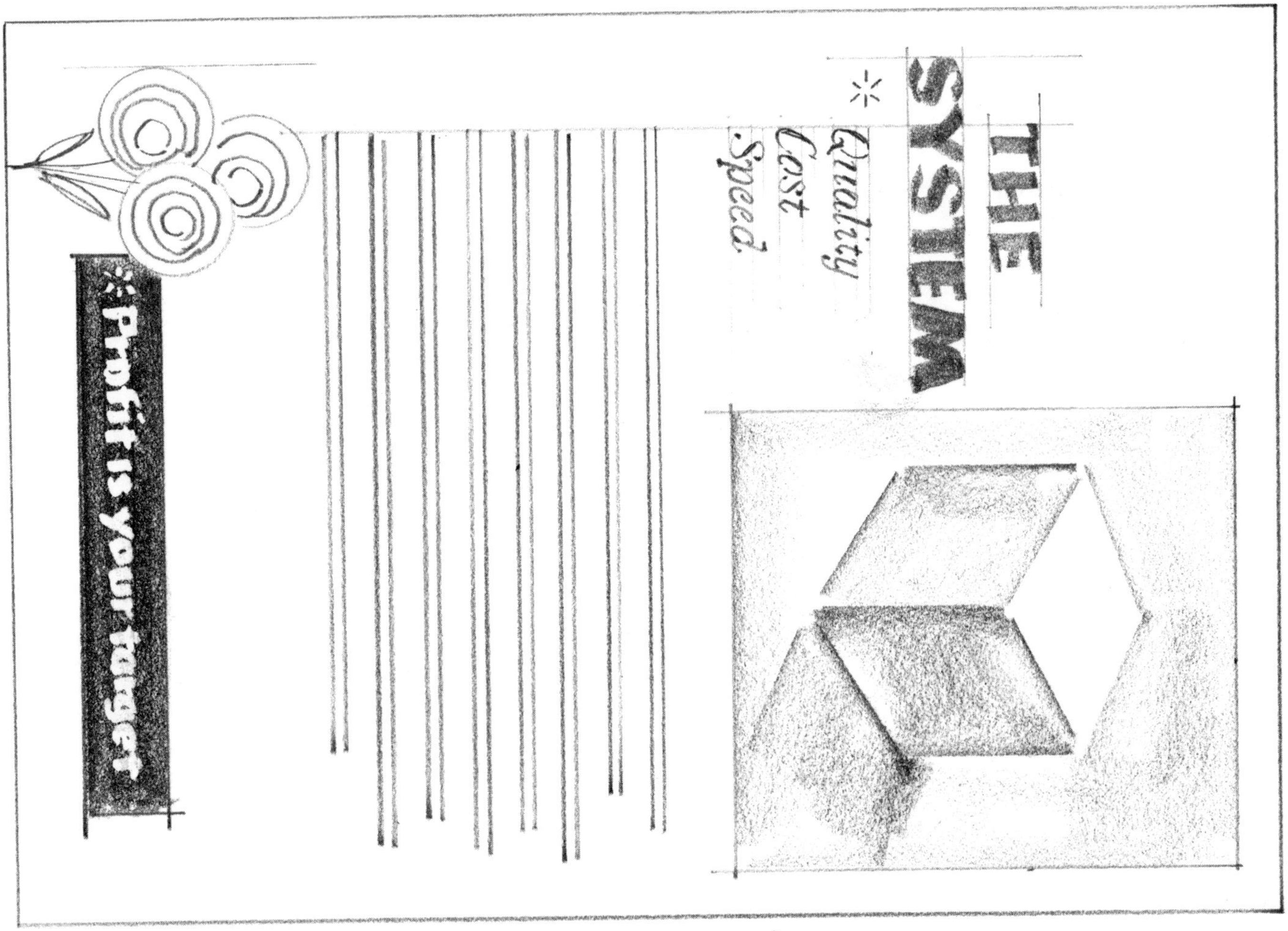

LAYOUT ON 6 3/16" x 8 1/2" SIZE SHEET

THE SYSTEM

Quality Cost Speed

TYPE GALLEY

Like many arts put to practical use, Paste-up can be seen as having three sides. It seems to be a balance of three forces. First it is an art form, and quality is the primary consideration. Yet, if it costs too much it becomes impractical. Further, to be practical, it must be available when it is needed. Many say that this might all be summed up by talking about cost, but that is too simple. Sometimes new ideas are the thing that is needid.

STAT

Like many arts put to practical use, Paste-up can be seen as having three sides.

EXTRA TYPE GALLEY

CLIP AND USE THIS PAGE FOR MATERIAL TO DO PASTE-UP

← PHOTO

*Profit is your target

← STAT

Quality PASTE-UP COMPANY

JOB TICKET

TO: ____________________
NAME

DATE ____________________
ADDRESS

____
CITY STATE ZIP

SHIPPING INSTRUCTIONS ____________________

____

____

JOB TICKET → JOB ____________________ JOB# ____________

CLIP AND USE THIS PAGE FOR MATERIAL TO DO PASTE-UP

Activistatom. DIVISION OF TECHNORESEARCH CORPORATION

ACTIVISTATOM COMPANY, • H20 PACIFICA PLACE, • TECHNOLOGY, EARTH AD2000, • (LUV) SUR-VIVE

Activistatom. DIVISION OF TECHNORESEARCH CORPORATION

ACTIVISTATOM COMPANY, • H20 PACIFICA PLACE, • TECHNOLOGY, EARTH AD2000, • (LUV) SUR-VIVE

ATOMIC ENERGIZER